SO-AVD-979

SIP

Understanding the Session Initiation Protocol

Second Edition

For a listing of recent titles in the *Artech House Telecommunications Library*, turn to the back of this book.

SIP

Understanding the Session Initiation Protocol

Second Edition

Alan B. Johnston

Artech House
Boston • London
www.artechhouse.com

Library of Congress Cataloging-in-Publication Data

A catalog record of this title is available from the Library of Congress

British Library Cataloguing in Publication Data
Johnston, Alan B.
    SIP : Understanding the Session Initiation Protocol. —2nd ed. — (Artech House
       telecommunications library)
    1. Computer network protocols
    I. Title
    004.6'2
    ISBN 1-58053-655-7

Cover design by Lisa Johnston

© 2004 ARTECH HOUSE, INC.
685 Canton Street
Norwood, MA 02062

All rights reserved. Printed and bound in the United States of America. No part of this book
may be reproduced or utilized in any form or by any means, electronic or mechanical, including
photocopying, recording, or by any information storage and retrieval system, without permission
in writing from the publisher.

    All terms mentioned in this book that are known to be trademarks or service marks have been
appropriately capitalized. Artech House cannot attest to the accuracy of this information. Use of
a term in this book should not be regarded as affecting the validity of any trademark or service
mark.

International Standard Book Number: 1-58053-655-7

10 9 8 7 6 5 4 3 2

To Lisa

Contents

Foreword to the First Edition

The Internet now challenges the close to $1 trillion world telecom industry. A renaissance in communications is taking place on the Internet. At its source are new communication protocols that would be impractical on the centralized control systems of circuit-switched networks used in telecommunications.

The Internet and the World Wide Web can be technically defined only by their protocols. Similarly, IP telephony and the wider family of IP communications are defined by several key protocols, most notably by the Session Initiation Protocol, or SIP.

The previously closed door of telecommunications is now wide open to web developers because of SIP and its relation to the web HTTP 1.1 protocol and the e-mail SMTP protocol. IP communications include voice/video, presence, instant messaging, mobility, conferencing, and even games. We believe many other communication areas are yet to be invented. The integration of all types of communications on the Internet may represent the next "killer application" and generate yet another wave of Internet growth.

As explained in this book, SIP is a close relative of the HTTP 1.1 and SMTP protocols. This represents a revolution in communications because it abandons the telecom signaling and control models developed for telephony over many years in favor of Internet and web-based protocols. Users and service providers obtain not only seamless integration of telephony and conferencing with many other World Wide Web and messaging applications, but also benefit from new forms of communications, such as presence and instant messaging.

Mobility can also be managed across various networks and devices using SIP. Location management is now under user control, so that incoming "calls" can be routed to any network and device that the called party may prefer. Users may even move across the globe to another service provider and maintain not only their URL "number", but also their personal tailored services and preferences. The end user gains control over all possible preferences, depending on various parameters such as who the other party is, what network he is on and what devices he is using, as well as time of day, subject, and other variables.

The new dimension in communications called "presence" enables users for the first time to indulge in "polite calling" by first sensing presence and preferences of the other party, before making a call. In its turn, presence can trigger location- and time-dependent user preferences. Users may want to be contacted in different ways, depending on their location and type of network access.

E-commerce will also benefit from IP communications. Extremely complex telecom applications, as found in call centers, have become even more complex when integrated with e-mail and web applications for e-commerce. Such applications, however, are quite straightforward to implement using SIP, due to its common structure with the web and e-mail. For example, both call routing and e-mail routing to agents—based on various criteria such as queue length, skill set, time of day, customer ID, the web page the customer is looking at, and customer history—can be reduced to simple XML scripts when using SIP and another IETF standard, the Call Processing Language (CPL). These examples are in no way exhaustive, but are mentioned here as a way of introduction.

This book starts with a short summary of the Internet, the World Wide Web, and its core protocols and addressing. Though familiar to many readers, these chapters provide useful focus on issues for the topics ahead. The introduction to SIP is made easy and understandable by examples that illustrate the protocol architecture and message details. Finally, in the core of the book, a methodical and complete explanation of SIP is provided. We refer the reader to the Table of Contents for a better overview and navigation through the topics.

Alan Johnston has made significant contributions toward the use of SIP for communications over the Internet. I had the privilege of watching Alan in meetings with some of the largest telecom vendors as he went methodically line by line over hundreds of call flows, which were then submitted as an Internet Draft to the Internet Engineering Task Force (IETF) and implemented in commercial systems. Alan combines in this book his expertise and methodical approach with page turning narrative and a discreet sense of humor.

I could not help reading the book manuscript page by page, since everything from Internet basics, protocols, and SIP itself is explained so well, in an attractive and concise manner.

Henry Sinnreich
Distinguished Member of Engineering
WorldCom
Richardson, Texas
July 2000

Preface to the Second Edition

Much has changed in the 2.5 years since the first edition of *SIP: Understanding the Session Initiation Protocol* was published. In 2001, SIP was a relatively unknown quantity, an upstart in the voice over IP (VoIP) and multimedia communications industry. Today, SIP is seen as the future of call signaling and telephony. It has been widely deployed by service providers and enterprises and is used casually every day by users of the dominant PC operating system. The full range of possibilities enabled by SIP is just now being glimpsed, and many more possibilities are yet to come.

One reason for this rapid acceptance is that SIP is an incredibly powerful call control protocol. It allows intelligent end points to implement the entire suite of telephony, Private Branch Exchange (PBX), Class, and Centrex services without a service provider, and without a controller or switch, for example.

The biggest driver for SIP on the Internet, however, has less to do with SIP's signaling and call control capabilities. Instead, it is due to the extensions of SIP that turn it into a powerful "rendezvous" protocol that leverages mobility and presence to allow users to communicate using different devices, modes, and services anywhere they are connected to the Internet. SIP applications provide support for presence—the ability to find out the status or location of a user without attempting to set up a session.

Another major change in the past few years is the adoption of wireless SIP to enable multimedia IP communications. As described in the chapter on wireless, SIP is now being used both in its standard form over 802.11 wireless networks and in planned commercial Third Generation Partnership Project (3GPP) rollouts in the coming years. SIP is ideally suited for this key application.

Since 2001 SIP has also grown in terms of the specification itself. Initially, SIP was described by a single RFC with a few related RFCs and a couple of RFC extensions. In Chapter 6 alone, more than 20 SIP-related RFCs are referenced. This book attempts to put all those documents together and provide a single reference for the protocol and all its extensions. Even SIP headers and responses that were standardized in the past but are now removed (deprecated) are listed in this text, providing useful context and background. Many others are discussed that are in the final stages of standardization prior to publication as RFCs, providing an up-to-date insider's view of the future of the protocol.

In closing, I again thank my colleagues in the Internet Engineering Task Force (IETF) and at MCI for all their contributions to the development of this protocol—it has been a privilege to be a part of a group of people that have created the SIP industry. Finally, I'd like to tip my hat to two of the key inventors of SIP who continue to develop and propel its implementation: Henning Schulzrinne and Jonathan Rosenberg.

Preface to the First Edition

When I began looking into the Session Initiation Protocol (SIP) in October 1998, I had prepared a list of a half dozen protocols relating to Voice over IP and Next Generation Networking. It was only a few days into my study that my list narrowed to just one: SIP. My background was in telecommunications, so I was familiar with the complex suite of protocols used for signaling in the Public Switched Telephone Network. It was readily apparent to me that SIP would be revolutionary in the telecommunications industry. Only a few weeks later I remember describing SIP to a colleague as the "SS7 of future telephony"—quite a bold statement for a protocol that almost no one had heard of, and that was not even yet a proposed standard!

Nearly 2 years later, I have continued to work almost exclusively with SIP since that day in my position with WorldCom, giving seminars and teaching the protocol to others. This book grew out of those seminars and my work on various Internet-Drafts.

This revolutionary protocol was also the discovery of a radical standards body—the Internet Engineering Task Force (IETF). Later, I attended my first IETF meeting, which was for me a career changing event. To interact with this dedicated band of engineers and developers, who have quietly taken the Internet from obscurity into one of the most important technological developments of the late 20th century, for the first time was truly exciting.

Just a few short years later, SIP has taken the telecommunications industry by storm. The industry press contains announcement after announcement of SIP product and service support from established vendor startups, and from established carriers. As each new group and company joins the dialog, the protocol has been able to adapt and grow without becoming unwieldy or overly complex. In the future, I believe that SIP, along with a TCP/IP stack, will find its way into practically every intelligent electronic device that has a need to communicate with the outside world.

With my telecommunications background, it is not surprising that I rely on telephone examples and analogies throughout this book to explain and illustrate SIP. This is also consistent with the probability that telecommunications is the first widely deployed use of the protocol. SIP stacks will soon be in multimedia PCs, laptops, palmtops, and in dedicated SIP telephones. The protocol will be used by telephone switches, gateways, wireless devices, and mobile phones. One of the key features of SIP, however, is its flexibility; as a result, the protocol is likely to be used in a whole host of applications that have little or nothing to do with telephony. Quite possibly one of these applications, such as instant messaging, may become the next "killer application" of the Internet. However, the operation and concepts of the protocol are unchanged regardless of the application, and the telephone analogies and examples are, I feel, easy to follow and comprehend.

The book begins with a discussion of the Internet, the IETF, and the Internet Multimedia Protocol Stack, of which SIP is a part. From there, the protocol is introduced by examples. Next, the elements of a SIP network are discussed, and the details of the protocol in terms of message types, headers, and response codes are covered. In order to make up a complete telephony system, related protocols, including Session Description Protocol (SDP) and Real-Time Transport Protocol (RTP), are covered. SIP is then compared to another signaling protocol, H.323, with the key advantages of SIP highlighted. Finally, the future direction of the evolution of the protocol is examined.

Two of the recurring themes of this book are the simplicity and stateless nature of the protocol. Simplicity is a hallmark of SIP due to its text-encoded, highly readable messages, and its simple transactional model with few exceptions and special conditions. Statelessness relates to the ability of SIP servers to store minimal (or no) information about the state or existence of a media session in a network. The ability of a SIP network to use stateless servers that do not need to record transactions, keep logs, fill and empty buffers, etc., is, I believe, a seminal step in the evolution of communications systems. I hope that these two themes become apparent as you read this book and learn about this exciting new protocol.

The text is filled with examples and sample SIP messages. I had to invent a whole set of IP addresses, domain names, and URLs. Please note that they are *all* fictional—do not try to send anything to them.

I would first like to thank the group of current and former engineers at WorldCom who shared their knowledge of this protocol and gave me the opportunity to author my first Internet-Draft document. I particularly thank Henry Sinnreich, Steve Donovan, Dean Willis, and Matt Cannon. I also thank

Robert Sparks, who I first met at the first seminar on SIP that I ever presented. Throughout the whole 3-hour session I kept wondering about the guy with the pony tail who seemed to know more than me about this brand new protocol! Robert and I have spent countless hours discussing fine points of the protocol. In addition, I would like to thank him for his expert review of this manuscript prior to publication—it is a better book due to his thoroughness and attention to detail. I also thank everyone on the IETF SIP list who has assisted me with the protocol and added to my understanding of it.

A special thanks to my wife Lisa for the terrific cover artwork and the cool figures throughout the book.

Finally, I thank my editor Jon Workman, the series editor and reviewer, and the whole team at Artech for helping me in this, my first adventure in publishing.

1

SIP and the Internet

The Session Initiation Protocol (SIP) is a new signaling, presence and instant messaging protocol developed to set up, modify, and tear down multimedia sessions, request and deliver presence and instant messages over the Internet [1]. This chapter covers some background for the understanding of the protocol. SIP was developed by the IETF as part of the Internet Multimedia Conferencing Architecture, and was designed to dovetail with other Internet protocols such as Transmission Control Protocol (TCP), Transmission Layer Security (TLS), User Datagram Protocol (UDP), Internet Protocol (IP), Domain Name System (DNS), and others. This organization and these related protocols will be briefly introduced. Related background topics such as Internet uniform resource indicators (URIs) and uniform resource locators (URLs), IP multicast routing, and Augmented Backus-Naur Format (ABNF) representations of protocol messages will also be covered.

1.1 Signaling Protocols

This book is about the Session Initiation Protocol. As the name implies, the protocol allows two end points to establish media sessions with each other. The main signaling functions of the protocol are as follows:

- Location of an end point;
- Contacting an end point to determine willingness to establish a session;
- Exchange of media information to allow session to be established;
- Modification of existing media sessions;

1

- Tear-down of existing media sessions.

SIP has also been extended to request and deliver presence information (on-line/off-line status and location information such as that contained in a "buddy" list) as well as instant message sessions. These functions include:

- Publishing and uploading of presence information;
- Requesting delivery of presence information;
- Presence and other event notification;
- Transporting of instant messages.

While some of the examples discuss SIP from a telephony perspective, there will be many nontelephony uses for SIP. SIP will likely be used to establish a whole set of session types that bear almost no resemblance to a telephone call.

1.2   The Internet Engineering Task Force

SIP was developed by the IETF. To quote "The Tao of the IETF" [2]: "The Internet Engineering Task Force is a loosely self-organized group of people who make technical and other contributions to the engineering and evolution of the Internet and its technologies." The two document types used within the IETF are Internet-Drafts (I-Ds) and Request for Comments (RFCs). I-Ds are the working documents of the group; anyone can author one on any topic and submit it to the IETF. There is no formal membership in the IETF; anyone can participate. Every I-D contains the following paragraph on the first page: "Internet-Drafts are documents valid for a maximum of six months and may be updated, replaced, or obsoleted by other documents at any time. It is inappropriate to use Internet-Drafts as reference material or to cite them other than as work in progress."

Internet standards are archived by the IETF as the Request for Comments series of numbered documents. As changes are made in a protocol, or new versions come out, a new RFC document with a new number is issued, which "obsoletes" the old RFC. Some I-Ds are cited in this book; I have tried, however, to restrict this to mature documents that are likely to become RFCs by the time this book is published. A standard begins its life as an I-D and then progresses to an RFC once there is consensus and there are working implementations of the protocol. Anyone with Internet access can download any I-D or RFC at no charge using the World Wide Web, ftp, or e-mail. Information on how to do so can be found on the IETF Web site: http://www.ietf.org.

The IETF is organized into working groups, which are chartered to work in a particular area and develop a protocol to solve that particular area. Each working group has its own archive and mailing list, which is where most of the work gets done. The IETF also meets three times per year.

1.3 A Brief History of SIP

SIP was originally developed by the IETF Multi-Party Multimedia Session Control Working Group, known as MMUSIC. Version 1.0 was submitted as an Internet-Draft in 1997. Significant changes were made to the protocol and resulted in a second version, version 2.0, which was submitted as an Internet-Draft in 1998. The protocol achieved Proposed Standard status in March 1999 and was published as RFC 2543 [3] in April 1999. In September 1999, the SIP working group was established by the IETF to meet the growing interest in the protocol. An Internet-Draft containing bug fixes and clarifications to SIP was submitted beginning in July 2000, referred to as RFC 2543 "bis". This document was eventually published as RFC 3261 [4], which obsoletes (or replaces) the original RFC 2543 specification. In addition, several SIP extension RFC documents have been published.

The popularity of SIP in the IETF has lead to the formation of other SIP-related working groups. The Session Initiation Protocol investigation (SIPPING) working group was formed to investigate applications of SIP, develop requirements for SIP extensions, and publish best current practice (BCP) documents about the use of SIP. This working group also publishes SIP event package RFCs. The SIP for Instant Messaging and Presence Leveraging Extensions (SIMPLE) working group was formed to standardize related protocols for presence and instant messaging applications. Other working groups that make use of SIP include PSTN and Internet Internetworking (PINT) working group and Service in the PSTN/IN requesting Internet Services (SPIRITS) working group.

To advance from Proposed Standard to Draft Standard, a protocol must have multiple independent interworking implementations and limited operational experience. Since the early days of RFC 2543, SIP interoperability test events, called SIPit (formerly called "bakeoffs"), have been held a few times per year. For the latest information about SIPit, visit the SIP Forum Web site: http://www.sipforum.org. (Note that the SIP Forum is a marketing/promotion organization for SIP and does not have any standardization function.) The final level, Standard, is achieved after operational success has been demonstrated [5]. With the documented interoperability from SIPit, SIP will move to Draft Standard status in the future.

SIP incorporates elements of two widely used Internet protocols: Hyper Text Transport Protocol (HTTP) used for Web browsing and Simple Mail Transport Protocol (SMTP) used for e-mail. From HTTP, SIP borrowed a client-server design and the use of URLs and URIs. From SMTP, SIP borrowed a text-encoding scheme and header style. For example, SIP reuses SMTP headers such as To, From, Date, and Subject. The interaction of SIP with other Internet protocols such as IP, TCP, UDP, and DNS will be described in the next section.

1.4   Internet Multimedia Protocol Stack

Figure 1.1 shows the four-layer Internet Multimedia Protocol stack. The layers shown and protocols identified will be discussed.

1.4.1   Physical Layer

The lowest layer is the physical and link layer, which could be an Ethernet local area network (LAN), a telephone line (V.90 or 56k modem) running Point-to-Point Protocol (PPP), or a digital subscriber line (DSL) running asynchronous transport mode (ATM), or even a wireless 802.11 network. This layer performs such functions as symbol exchange, frame synchronization, and physical interface specification.

1.4.2   Internet Layer

The next layer in Figure 1.1 is the Internet layer. Internet Protocol (IP) [6] is used at this layer to route a packet across the network using the destination IP address. IP is a connectionless, best-effort packet delivery protocol. IP packets can be lost, delayed, or received out of sequence. Each packet is routed on its own, using the IP header appended to the physical packet. Most IP address examples in this book use the older version of IP, version 4 (IPv4). IPv4

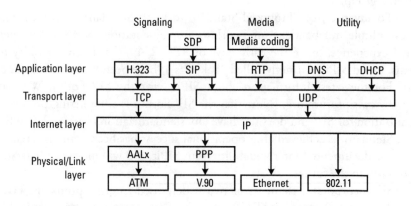

Figure 1.1 The Internet Multimedia Protocol stack.

addresses are four octets long, usually written in so-called "dotted decimal" notation (for example, 207.134.3.5). Between each of the dots is a decimal number between 0 and 255. At the IP layer, packets are not acknowledged. A checksum is calculated to detect corruption in the IP header, which could cause a packet to become misrouted. Corruption or errors in the IP payload, however, are not detected; a higher layer must perform this function if necessary. IP uses a single-octet protocol number in the packet header to identify the transport layer protocol that should receive the packet.

IP version 6 (IPv6) [7] was developed by the IETF as a replacement for IPv4. It has been slowly gaining support and is supported now by a number of operating systems. The biggest initial networks of IPv6 will likely be the wireless telephony carriers who need the most important advantage of IPv6 over IPv4—that is, a much enlarged addressing space. IPv6 increases the addressing space from 32 bits in IPv4 to 128 bits, providing for over 4 billion IPv6 addresses. There are many other improvements in IPv6 over IPv4, including security. An IPv6 address is typically written as a sequence of eight hexadecimal numbers separated by colons. For example, 0:0:0:0:aaaa:bbbb:cccc:dddd is an IPv6 address written in this format. It is also common to drop sequences of zeros with a single double colon. The same address can then be written as ::aaaa:bbbb:cccc:dddd.

IP addresses used over the public Internet are assigned in blocks by the Internet Assigned Number Association (IANA). As a result of this centralized assignment, IP addresses are globally unique. This enables a packet to be routed across the public Internet using only the destination IP address. Various protocols are used to route packets over an IP network, but they are outside of the scope of this book. Subnetting and other aspects of the structure of IP addresses are also not covered here. There are other excellent sources [8] that cover the entire suite of TCP/IP protocols in more detail.

1.4.3 Transport Layer

The next layer shown in Figure 1.1 is the transport layer. It uses a two-octet port number from the application layer to deliver the datagram or segment to the correct application layer protocol at the destination IP address. Some port numbers are dedicated to particular protocols—these ports are called "well-known" port numbers. For example, HTTP uses the well-known port number of 80, while SIP uses the well-known port number of 5060 or 5061. Other port numbers can be used for any protocol, and they are assigned dynamically from a pool of available numbers. These so-called "ephemeral" port numbers are usually in the range 49152 to 65535. There are three commonly used transport layer protocols, Transmission Control Protocol, Transmission Layer Security, and User Datagram Protocol, which are described in the following sections.

1.4.3.1   TCP

TCP [9] provides reliable, connection-oriented transport over IP. TCP uses sequence numbers and positive acknowledgments to ensure that each block of data, called a segment, has been received. Lost segments are retransmitted until they are successfully received. Figure 1.2 shows the message exchange to establish and tear down a TCP connection. A TCP server "listens" on a well-known port for a TCP request. The TCP client sends a SYN (synchronization) message to open the connection. The SYN message contains the initial sequence number the client will use during the connection. The server responds with a SYN message containing its own initial sequence number, and an acknowledgment number, indicating that it received the SYN from the client. The client completes the three-way handshake with an ACK or a DATA packet with the AK flag set to the server acknowledging the server's sequence number. Now that the connection is open, either client or server can send data in DATA packets called segments.

Each time a sender transmits a segment, it starts a timer. If a segment is lost in transmission, this timer will expire. Deducing a lost segment, the sender will resend the segment until it receives the acknowledgment. The FIN message closes the TCP connection. The sequence of four messages shown in Figure 1.2 closes the connection. The ephemeral port numbers used in the connection are then free to be used in establishing other connections. TCP also has built-in mechanisms for flow control. During the SYN processing, a window size

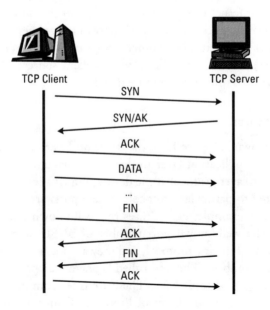

Figure 1.2 Opening and closing a TCP connection.

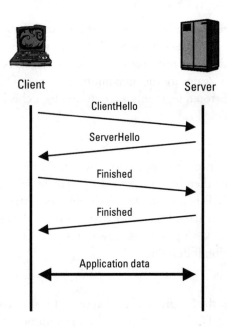

Figure 1.3 Opening of TLS connections.

Another advantage is that SCTP avoids the so-called "head of line blocking" problem of TCP. This is a common TCP problem in which a dropped segment with a large window causes the entire window's worth of messages to wait in a buffer (i.e., be blocked) until the dropped segment is retransmitted.

SCTP also supports multihoming, so if one of a pair of load balancing servers fails, the other can immediately begin receiving the messages without even requiring a DNS or other database lookup.

SCTP is a true layer 2 transport protocol that requires operating system level support to be used, which will initially delay its use in the Internet. Also, note that the advantages of SCTP over TCP only occur during packet loss. In a zero loss network, the performance of the two is identical.

1.4.4   Application Layer

The top layer shown in Figure 1.1 is the application layer. This includes signaling protocols such as SIP and media transport protocols such as Real-time Transport Protocol (RTP), which is introduced in Section 7.2. Figure 1.1 includes H.323, introduced in Chapter 8, which is an alternative signaling protocol to SIP developed by the International Telecommunication Union (ITU). Session Description Protocol (SDP), described in Section 7.1, is shown above SIP in the protocol stack because it is carried in a SIP message body. HTTP,

SMTP, FTP, and Telnet are all examples of application layer protocols. Because SIP can use any transport protocol, it is shown interacting with both TCP, TLS, and UDP in Figure 1.1. The use of TCP, TLS, SCTP, and UDP transport for SIP will be discussed in the next chapter.

1.5 Utility Applications

Two utility applications are also shown in Figure 1.1 as users of UDP. The most common use of the DNS (well-known port number 53) is to resolve a symbolic name (such as domain.com, which is easy to remember) into an IP address (which is required by IP to route the packet). Also shown is the Dynamic Host Configuration Protocol (DHCP). DHCP allows an IP device to download configuration information upon initialization. Common fields include a dynamically assigned IP address, DNS addresses, subnet masks, maximum transmission unit (MTU), or maximum packet size, and server addresses for mail and Web browsing. Figure 1.4 shows the layer interaction for processing a request. At the top, a URL from the user layer is input to the application layer. URLs, described

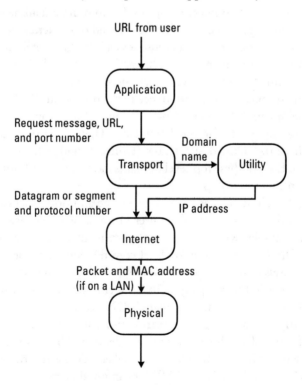

Figure 1.4 Request processing in the Internet Protocol stack.

later in this chapter, are names used to represent resources, hosts, or files on the Internet. The application passes the generated request (for example, a HTTP GET request, which requests a Web page download), the URL, and the port number to the transport layer. The transport layer uses a utility to resolve the domain name extracted from the URL into an IP address. The IP address, datagram (or segment, depending on the transport layer protocol used), and protocol number identifying the transport protocol are then passed to the Internet layer. The Internet layer then passes the packet to the physical layer along with a media access control (MAC) address for routing, in the case of a LAN. A response is processed by reversing the above steps. A response received at the physical layer flows back up the layers, with the header information being stripped off and the response data passed upwards towards the user. The main difference is that no utility is used in response processing.

1.6   DNS and IP Addresses

Domain Name Service [13] is used in the Internet to map a symbolic name (such as www.amazon.com) to an IP address (such as 100.101.102.103). DNS is also used to obtain information needed to route e-mail messages and, in the future, SIP messages. The use of names instead of numerical addresses is one of the Internet's greatest strengths because it gives the Internet a human, friendly feel. Domain names are organized in a hierarchy. Each level of the name is separated by a *dot*, with the highest level domain on the right side. (Note that the dots in a domain name have no correspondence to the dots in an IP address written in dotted decimal notation.) General top-level domains are shown in Table 1.1 (see http://www.icann.org/tlds for the latest list). There is also a set of country domains such as: us (United States), uk (United Kingdom), ca (Canada), and au (Australia). Each of these top-level domains has just one authority that assigns that domain to a user or group.

Once a domain name has been assigned, the authority places a link in their DNS server to the DNS server of the user or group who has been assigned the domain. For example, when company.com is allocated to a company, the authoritative DNS server for the top-level com domain entry for company contains the IP address of the company's DNS server(s). A name can then be further qualified by entries in the company's DNS server to point to individual servers in their network. For example, the company's DNS server may contain entries for www.company.com, ftp.company.com, and smtp.company.com. A number of types of DNS record types are defined. The DNS records used to resolve a host name into an IP address are called address records, or A records. Other types of records include CNAME (or canonical name or alias records), MX (or mail exchange records), SRV (or Service records, used by SIP and other

Table 1.1
Internet Top-Level Domains

Domain	Description
com	Company
net	Network
int	Internet
org	Not-for-profit organization
edu	University or college
gov	U.S. government
mil	U.S. military
arpa	ARPAnet
info	Information
biz	Business
museum	Museum
name	Name
pro	Professional
aero	Air transport industry
coop	Cooperatives

protocols), and TXT (or free-form text records). Another type of DNS record is a PTR, or pointer record, used for reverse lookups. Reverse lookups are used to map an IP address back to a domain name. These records can be used, for example, in generating server logs that show not only the IP addresses of clients served, but also their domain name. Web browsing provides an example of the use of the DNS system. Another type of DNS record is known as a Naming Authority Pointer (NAPTR) record that can be used by a protocol known as ENUM [14] to map global telephone numbers into Internet URLs.

When a user types in a Web address, such as www.artech-house.com, the name must be resolved to an IP address before the browser can send the request for the index Web page from the Artech House Web server. The Web browser first launches a DNS query to the IP address for its DNS server, which has been manually configured or set up using DHCP. If the DNS server happens to have the name's A record stored locally (cached) from a recent query, it will return the IP address. If not, the DNS root server will then be queried to locate the authoritative DNS server for Artech House, which must contain the A records for the artech-house.com domain. The HTTP GET request is then sent to that IP address, and the Web browsing session begins. There is only one authoritative DNS server for a domain, and it is operated by the owner of the domain name. Due to a very efficient caching scheme built into DNS, a DNS request often does not have to route all the way to this server. DNS is also used by an SMTP server to deliver an e-mail message. An SMTP server with an e-mail message to deliver initiates a DNS request for the MX record of the domain name in the destination e-mail address. The response to the request allows the SMTP server to contact the destination SMTP server and transfer the message. A similar process has been proposed for locating a SIP server using SRV, or service, DNS records.

1.7 URLs and URIs

Uniform resource locators [15] are names used to represent addresses or locations in the Internet. URLs are designed to encompass a wide range of protocols and resource types in the Internet. The basic form of a URL is `scheme:specifier`—for example, `http://www.artech-house.com/search/search.html`. The token `http` identifies the scheme or protocol to be used, in this case HTTP. The specifier follows the ":" and contains a domain name (`www.artechhouse.com`), which can be resolved into an IP address and a file name (`/search/search.html`). URLs can also contain additional parameters or qualifiers relating to transport. For example, `telnet://host.company.com:24` indicates that the Telnet Protocol should be used to access `host.company.com` using port 24. New schemes for URLs for new protocols are easily constructed, and dozens have been defined, such as `mailto`, `tel`, and `https`. The `sip` and `sips` schemes will be introduced in Section 2.2 and are described in detail in Section 4.2.

Most protocols reference URLs, but with SIP we mainly reference URIs. This is due to the mobility aspects of SIP, which means that a particular address (URI) is not tied to a single physical device but instead is a logical entity that may move around and change its location in the Internet. However, the terms URL and URI are often used almost interchangeably in other contexts.

1.8 Multicast

In normal Internet packet routing, or unicast routing, a packet is routed to a single destination. In multicast routing, a single packet is routed to a set of destinations. Single LAN segments running a protocol such as Ethernet offer the capability for packet broadcast, where a packet is sent to every node on the network. Scaling this to a larger network with routers is a recipe for disaster, as broadcast traffic can quickly cause congestion. An alternative approach for this type of packet distribution is to use a packet reflector that receives packets and forwards copies to all destinations that are members of a broadcast group. This also can cause congestion in the form of a "packet storm" [16]. For a number of years, the Internet Multicast Backbone Network (MBONE), an overlay of the public Internet, has used multicast routing for high-bandwidth broadcast sessions. Participants who wish to join a multicast session send a request to join the session to their local MBONE router. That router will then begin to broadcast the multicast session on that LAN segment. Additional requests to join the session from others in the same LAN segment will result in no additional multicast packets being sent, since the packets are already being broadcast. If the router is not aware of any multicast participants on its segment, it will not forward any of the packets. Routing of multicast packets between routers uses special multicast routing protocols to ensure that packet traffic on the backbone is kept to a minimum. Multicast Internet addresses are reserved in the range 224.0.0.0 to 239.255.255.255. Multicast transport is always UDP, since the handshake and acknowledgments of TCP are not possible. Certain addresses have been defined for certain protocols and applications. The scope or extent of a multicast session can be limited using the time-to-live (TTL) field in the IP header. This field is decremented by each router that forwards the packet, which limits the number of hops the packet takes. SIP support for multicast will be discussed in Section 3.10. Multicast is slowly becoming a part of the public Internet as service providers begin supporting it.

1.9 ABNF Representation

The meta-language Augmented Backus-Naur Format [17] is used throughout RFC 3261 [1] to describe the syntax of SIP, as well as other Internet protocols. An example construct used to describe a SIP message is as follows:

```
SIP-message = Request / Response
```

This is read: A SIP message is either a request or a response. SIP-message on the left side of the "equals" sign represents what is being defined. The right side of the "equals" sign contains the definition. The "/" is used to mean logical OR (note: older versions of ABNF used "|" in place of "/" in grammars). Next, Request and Response are defined in a similar manner using ABNF:

```
Request = Request-Line *( message-header ) CRLF [ message-body ]
```

Request-Line will be defined in another ABNF statement. Elements enclosed in ( ) are treated as a single element. The "*" means the element may be repeated, separated by at least one space. The minimum and maximum numbers can be represented as x*y, which means a minimum of x and maximum of y. Since the default values are 0 and infinity, a solitary "*" (as in this example) indicates any number is allowed, including none. CRLF is defined as a carriage return line feed, or the ASCII characters that are written in Internet hexadecimal notation as 0x10 and 0x13. Other common ABNF representations include SP for space (ASCII 0x32). A message body is optional in a Request, and is enclosed in square brackets [ ] to indicate this. Comments in the ABNF begin with a semicolon ";" and continue to the end of the line. Lines continue the same ABNF definition when they are indented. Tokens are defined in ABNF as any set of characters besides control characters and separators. Display names and other components of a SIP header that are not used by the protocol are considered tokens; they are simply parsed and ignored. For example:

```
transport = "UDP" / "TCP" / "TLS" / "SCTP"
```

This ABNF defines four possible tokens, which may be used in a transport parameter. In this text, few references to ABNF will be made. Instead, SIP messages and elements will be introduced by description and example rather than by using ABNF. The ABNF for SIP is in Section 25 of RFC 3261.

References

[1]   Leiner, B., et al., "A Brief History of the Internet," The Internet Society, http://www.isoc.org/internet/history/brief.html.

[2]   Malkin, G., and the IETF Secretariat, "The Tao of the IETF—A Guide for New Attendees of the Internet Engineering Task Force," http://www.ietf.org.tao.html.

[3]   Handley, M., et al., "SIP: Session Initiation Protocol," RFC 2543, 1999.

[4]   Rosenberg, J., et al., "SIP: Session Initiation Protocol," RFC 3261, 2002.

[5]   Bradner, S., "The Internet Standards Process: Revision 3," RFC 2026, 1996.

[6]   "Internet Protocol," RFC 791, 1981.

[7]   Deering, S., and R. Hinden, "Internet Protocol, Version 6 (IPv6) Specification," RFC 1883, 1995.

[8]   Wilder, F., *A Guide to the TCP/IP Protocol Suite*, Norwood, MA: Artech House, 1998.

[9]   "Transmision Control Protocol," RFC 793, 1981.

[10] Postal, J., "User Datagram Protocol," RFC 768, 1980.

[11] Dierks, T., et al., "The TLS Protocol Version 1.0," RFC 2246, 1999.

[12] Stewart, R., et al., "Stream Control Transmission Protocol," RFC 2960, 1999.

[13] Manning, B., "DNS NSPA RRs," RFC 1348, 1992.

[14] Falstrom, P., "E.164 Number and DNS," RFC 2916, 2000.

[15] Berners-Lee, T., L. Masintes, and M. McCahill, "Uniform Resource Locators," RFC 1738, 1994.

[16] Hersent, O., D. Gurle, and J. Petit, *IP Telephony Packet-Based Multimedia Communications Systems*, Reading, MA: Addison-Wesley, 2000, Chapter 8.

[17] Crocker, D., "Standard for the Format of ARPA Internet Text Messages," RFC 822, 1982.

2

Introduction to SIP

Often the best way to learn a protocol is to look at examples of its use. While the terminology, structures, and format of a new protocol can be confusing at first read, an example message flow can give a quick grasp of some of the key concepts of a protocol. The example message exchanges in this chapter will introduce SIP as defined by RFC 3261 [1].

The first example shows the basic message exchange between two SIP devices to establish and tear down a session. The second example shows the message exchange when a SIP proxy server is used. The third example shows SIP registration. The fourth example shows a SIP presence and instant message example. The chapter concludes with a discussion of SIP message transmission using UDP, TCP, TLS, and SCTP.

The examples will be introduced using call flow diagrams between a called and calling party, along with the details of each message. Each arrow in the figures represents a SIP message, with the arrowhead indicating the direction of transmission. The thick lines in the figures indicate the media stream. In these examples, the media will be assumed to be RTP [2] packets containing audio, but it could be another protocol. Details of RTP are covered in Section 7.2.

2.1 A Simple Session Establishment Example

Figure 2.1 shows the SIP message exchange between two SIP-enabled devices. The two devices could be SIP phones, hand-helds, palmtops, or cell phones. It is assumed that both devices are connected to an IP network such as the Internet and know each other's IP address.

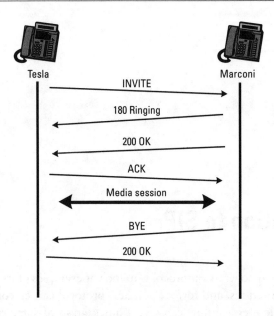

Figure 2.1 A simple SIP session establishment example.

The calling party, Tesla, begins the message exchange by sending a SIP INVITE message to the called party, Marconi. The INVITE contains the details of the type of session or call that is requested. It could be a simple voice (audio) session, a multimedia session such as a video conference, or it could be a gaming session.

The INVITE message contains the following fields:

```
INVITE sip:marconi@radio.org SIP/2.0
Via: SIP/2.0/UDP lab.high-voltage.org:5060;branch=z9hG4bKfw19b
Max-Forwards: 70
To: G. Marconi <sip:Marconi@radio.org>
From: Nikola Tesla <sip:n.tesla@high-voltage.org>;tag=76341
Call-ID: 123456789@lab.high-voltage.org
CSeq: 1 INVITE
Subject: About That Power Outage...
Contact: <sip:n.tesla@lab.high-voltage.org>
Content-Type: application/sdp
Content-Length: 158

v=0
o=Tesla 2890844526 2890844526 IN IP4 lab.high-voltage.org
s=Phone Call
c=IN IP4 100.101.102.103
t=0 0
m=audio 49170 RTP/AVP 0
a=rtpmap:0 PCMU/8000
```

Since SIP is a text-encoded protocol, this is actually what the SIP message would look like "on the wire" as a UDP datagram being transported over, for example, Ethernet.

The fields listed in the INVITE message are called header fields. They have the form Header: Value CRLF. The first line of the request message, called the start line, lists the method, which is INVITE, the Request-URI, then the SIP version number (2.0), all separated by spaces. Each line of a SIP message is terminated by a CRLF. The Request-URI is a special form of SIP URI and indicates the resource to which the request is being sent, also known as the request target. SIP URIs are discussed in more detail in later sections.

The first header field following the start line shown is a Via header field. Each SIP device that originates or forwards a SIP message stamps its own address in a Via header field, usually written as a host name that can be resolved into an IP address using a DNS query. The Via header field contains the SIP version number (2.0), a "/", then UDP for UDP transport, a space, then the hostname or address, a colon, then a port number, in this example the "well-known" SIP port number 5060. Transport of SIP using TCP, UDP, TLS, and SCTP and the use of port numbers are covered later in this chapter. The branch parameter is a transaction identifier. Responses relating to this request can be correlated because they will contain this same transaction identifier.

The next header field shown is the Max-Forwards header field, which is initialized to some large integer and decremented by each SIP server, which receives and forwards the request, providing simple loop detection.

The next header fields are the To and From header fields, which show the originator and destination of the SIP request. When a name label is used, as in this example, the SIP URI is enclosed in brackets and used for routing the request. The name label could be displayed during alerting, for example, but is not used by the protocol.

The Call-ID header field is an identifier used to keep track of a particular SIP session. The originator of the request creates a locally unique string, then usually adds an "@" and its host name to make it globally unique. In addition to the Call-ID, each party in the session also contributes a random identifier, unique for each call. These identifiers, called tags, are included in the To and From header fields as the session is established. The initial INVITE shown contains a From tag but no To tag.

The user agent that generates the initial INVITE to establish the session generates the unique Call-ID and From tag. In the response to the INVITE, the user agent answering the request will generate the To tag. The combination of the local tag (contained in the From header field), remote tag (contained in the To header field), and the Call-ID uniquely identifies the established session, known as a "dialog." This dialog identifier is used by both parties to identify this call because they could have multiple calls set up between them. Subsequent requests within the established session will use this dialog identifier, as will be shown in the following examples.

The next header field shown is the CSeq, or command sequence. It contains a number, followed by the method name, INVITE in this case. This number is incremented for each new request sent. In this example, the command sequence number is initialized to 1, but it could start at another integer value.

The Via header fields plus the Max-Forwards, To, From, Call-ID, and CSeq header fields represent the minimum required header field set in any SIP request message. Other header fields can be included as optional additional information, or information needed for a specific request type. A Contact header field is also required in this INVITE message, which contains the SIP URI of Tesla's communication device, known as a user agent (UA); this URI can be used to route messages directly to Tesla. The optional Subject header field is present in this example. It is not used by the protocol, but could be displayed during alerting to aid the called party in deciding whether to accept the call. The same sort of useful prioritization and screening commonly performed using the Subject and From header fields in an e-mail message is also possible with a SIP INVITE request. Additional header fields are present in this INVITE message, which contain the media information necessary to set up the call.

The Content-Type and Content-Length header fields indicate that the message body is SDP [3] and contains 158 octets of data. The basis for the octet count of 158 is shown in Table 2.1, where the CR LF at the end of each line is shown as a ©® and the octet count for each line is shown on the right-hand side. A blank line separates the message body from the header field list, which ends with the Content-Length header field. In this case, there are seven lines of SDP data describing the media attributes that the caller Tesla desires for the call. This media information is needed because SIP makes no assumptions about the type of media session to be established—the caller must specify exactly what type of session (audio, video, gaming) that he wishes to establish. The SDP field names are listed in Table 2.2, and will be discussed detail in Section 7.1, but a quick review of the lines shows the basic information necessary to establish a session.

Table 2.2 includes the:

- Connection IP address (100.101.102.103);
- Media format (audio);
- Port number (49170);
- Media transport protocol (RTP);
- Media encoding (PCM μ Law);

Table 2.1
Content-Length Calculation Example

LINE	TOTAL
v=0©®	05
o=Tesla 2890844526 2890844526 IN IP4 lab.high-voltage.org©®	59
s=Phone Call©®	14
c=IN IP4 100.101.102.103©®	26
t=0 0©®	07
m=audio 49170 RTP/AVP 0©®	25
a=rtpmap:0 PCMU/8000©®	22
	158

Table 2.2
SDP Data from Example

SDP Parameter	Parameter Name
v=0	Version number
o=Tesla 2890844526 2890844526 IN IP4 lab.high-voltage.org	Origin containing name
s=Phone Call	Subject
c=IN IP4 100.101.102.103	Connection
t=0 0	Time
m=audio 49170 RTP/AVP 0	Media
a=rtpmap:0 PCMU/8000	Attributes

- Sampling rate (8,000 Hz).

INVITE is an example of a SIP request message. There are five other methods or types of SIP requests currently defined in the SIP specification RFC 3261 and others in extension RFCs. The next message in Figure 2.1 is a 180 Ringing message sent in response to the INVITE. This message indicates that the called party Marconi has received the INVITE and that alerting is taking place. The alerting could be ringing a phone, flashing a message on a screen, or any other method of attracting the attention of the called party, Marconi.

The 180 Ringing is an example of a SIP response message. Responses are numerical and are classified by the first digit of the number. A 180 response is an "informational class" response, identified by the first digit being a 1. Informational responses are used to convey noncritical information about the progress of the call. Many SIP response codes were based on HTTP version 1.1 response codes with some extensions and additions. Anyone who has ever browsed the World Wide Web has likely received a "404 Not Found" response from a Web server when a requested page was not found. 404 Not Found is also a valid SIP "client error class" response in a request to an unknown user. The other classes of SIP responses are covered in Chapter 5.

The response code number in SIP alone determines the way the response is interpreted by the server or the user. The reason phrase, Ringing in this case, is suggested in the standard, but any text can be used to convey more information. For instance, 180 Hold your horses, I'm trying to wake him up! is a perfectly valid SIP response and has the same meaning as a 180 Ringing response.

The 180 Ringing response has the following structure:

```
SIP/2.0 180 Ringing
Via: SIP/2.0/UDP lab.high-voltage.org:5060;branch=z9hG4bKfw19b
  ;received=100.101.102.103
To: G. Marconi <sip:marconi@radio.org>;tag=a53e42
From: Nikola Tesla <sip:n.tesla@high-voltage.org>>;tag=76341
Call-ID: 123456789@lab.high-voltage.org
CSeq: 1 INVITE
Contact: <sip:marconi@tower.radio.org>
Content-Length: 0
```

The message was created by copying many of the header fields from the INVITE message, including the Via, To, From, Call-ID, and CSeq, then adding a response start line containing the SIP version number, the response code, and the reason phrase. This approach simplifies the message processing for responses.

The Via header field contains the original branch parameter but also has an additional received parameter. This parameter contains the literal IP address that the request was received from (100.101.102.103), which typically is the same address that the URI in the Via resolves using DNS (lab.high-voltage.org).

Note that the To and From header fields are not reversed in the response message as one might expect them to be. Even though this message is sent to Marconi from Tesla, the header fields read the opposite. This is because the To and From header fields in SIP are defined to indicate the direction of the request, not the direction of the message. Since Tesla initiated this request, all responses will read To: Marconi From: Tesla.

The To header field now contains a tag that was generated by Marconi. All future requests and responses in this session or dialog will contain both the tag generated by Tesla and the tag generated by Marconi.

The response also contains a Contact header field, which contains an address at which Marconi can be contacted directly once the session is established.

When the called party Marconi decides to accept the call (i.e., the phone is answered), a 200 OK response is sent. This response also indicates that the type of media session proposed by the caller is acceptable. The 200 OK is an example of a "success class" response. The 200 OK message body contains Marconi's media information:

```
SIP/2.0 200 OK
Via: SIP/2.0/UDP lab.high-voltage.org:5060;branch=z9hG4bKfw19b
  ;received=100.101.102.103
To: G. Marconi <sip:marconi@radio.org>;tag=a53e42
From: Nikola Tesla <sip:n.tesla@high-voltage.org>;tag=76341
Call-ID: 123456789@lab.high-voltage.org
CSeq: 1 INVITE
Contact: <sip:marconi@tower.radio.org>
Content-Type: application/sdp
Content-Length: 155

v=0
o=Marconi 2890844528 2890844528 IN IP4 tower.radio.org
s=Phone Call
c=IN IP4 200.201.202.203
t=0 0
m=audio 60000 RTP/AVP 0
a=rtpmap:0 PCMU/8000
```

This response is constructed the same way as the 180 Ringing response and contains the same To tag and Contact URI. The media capabilities, however, must be communicated in a SDP message body added to the response. From the same SDP fields as Table 2.2, the SDP contains:

- End-point IP address (200.201.202.203);
- Media format (audio);
- Port number (60000);
- Media transport protocol (RTP);
- Media encoding (PCM μ-Law);
- Sampling rate (8,000 Hz).

The final step is to confirm the media session with an "acknowledgment" request. The confirmation means that Tesla has received successfully Marconi's response. This exchange of media information allows the media session to be established using another protocol, RTP in this example.

```
ACK sip:marconi@tower.radio.org SIP/2.0
Via: SIP/2.0/UDP lab.high-voltage.org:5060;branch=z9hG4bK321g
Max-Forwards: 70
To: G. Marconi <sip:marconi@radio.org>;tag=a53e42
From: Nikola Tesla <sip:n.tesla@high-voltage.org>;tag=76341
Call-ID: 123456789@lab.high-voltage.org
CSeq: 1 ACK
Content-Length: 0
```

The command sequence, CSeq, has the same number as the INVITE, but the method is set to ACK. At this point, the media session begins using the media information carried in the SIP messages. The media session takes place using another protocol, typically RTP. The branch parameter in the Via header field contains a new transaction identifier than the INVITE, since an ACK sent to acknowledge a 200 OK is considered a separate transaction.

This message exchange shows that SIP is an end-to-end signaling protocol. A SIP network, or SIP server is not required for the protocol to be used. Two end points running a SIP protocol stack and knowing each other's IP addresses can use SIP to set up a media session between them. Although less obvious, this example also shows the client-server nature of the SIP protocol. When Tesla originates the INVITE request, he is acting as a SIP client. When Marconi responds to the request, he is acting as a SIP server. After the media session is established, Marconi originates the BYE request and acts as the SIP client, while Tesla acts as the SIP server when he responds. This is why a SIP-enabled device must contain both SIP server and SIP client software—during a typical session, both are needed. This is quite different from other client-server Internet protocols such as HTTP or FTP. The Web browser is always an HTTP client, and the Web server is always an HTTP server, and similarly for FTP. In SIP, an end point will switch back and forth during a session between being a client and a server.

In Figure 2.1, a BYE request is sent by Marconi to terminate the media session:

```
BYE sip:n.tesla@lab.high-voltage.org SIP/2.0
Via: SIP/2.0/UDP tower.radio.org:5060;branch=z9hG4bK392kf
Max-Forwards: 70
To: Nikola Tesla <sip:n.tesla@high-voltage.org>;tag=76341
From: G. Marconi <sip:marconi@radio.org>;tag=a53e42
Call-ID: 123456789@lab.high-voltage.org
CSeq: 1 BYE
Content-Length: 0
```

The Via header field in this example is populated with Marconi's host address and contains a new transaction identifier since the BYE is considered a separate transaction from the INVITE or ACK transactions shown previously. The To and From header fields reflect that this request is originated by Marconi, as they are reversed from the messages in the previous transaction. Tesla, however, is able to identify the dialog using the presence of the same local and remote tags and Call-ID as the INVITE, and tear down the correct media session.

Notice that all the branch IDs shown in the example so far begin with the string z9hG4bK. This is a special string that indicates that the branch ID has been calculated using strict rules defined in RFC 3261 and is as a result usable as a transaction identifier.[1]

The confirmation response to the BYE is a 200 OK:

```
SIP/2.0 200 OK
Via: SIP/2.0/UDP tower.radio.org:5060;branch=z9hG4bK392kf
  ;received=200.201.202.203
To: Nikola Tesla <sip:n.tesla@high-voltage.org>;tag=76341
From: G. Marconi <sip:marconi@radio.org>;tag=a53e42
Call-ID: 123456789@lab.high-voltage.org
CSeq: 1 BYE
Content-Length: 0
```

The response echoes the CSeq of the original request: 1 BYE.

2.2 SIP Call with Proxy Server

In the SIP message exchange of Figure 2.1, Tesla knew the IP address of Marconi and was able to send the INVITE directly to that address. This will not be the case in general—an IP address cannot be used like a telephone number. One reason is that IP addresses are often dynamically assigned due to the shortage of IPv4 addresses. For example, when a PC dials in to an Internet service provider (ISP) modem bank, an IP address is assigned using DHCP to the PC from a pool of available addresses allocated to the ISP. For the duration of the session, the IP address does not change, but it is different for each dial-in session. Even for an "always on" Internet connection such as a DSL line, a different IP address can be assigned after each reboot of the PC. Also, an IP address does not uniquely identify a user, but identifies a node on a particular physical IP network. You have one IP address at your office, another at home, and still another when you log on remotely when you travel. Ideally, there would be one address

---

1. This string is needed because branch IDs generated by user agents prior to RFC 3261 may have constructed branch IDs which are not suitable as transaction identifiers. In this case, a client must construct its own transaction identifier using the To tag, From tag, Call-ID, and CSeq.

that would identify you wherever you are. In fact, there is an Internet protocol that does exactly that, with e-mail. SMTP uses a host or system independent name (an e-mail address) that does not correspond to a particular IP address. It allows e-mail messages to reach you regardless of what your IP address is and where you are logged on to the Internet.

In addition, a request routed using only IP addresses will reach only one end point—only one device. Since communication is typically user-to-user instead of device-to-device, a more useful addressing scheme would allow a particular user to call another particular user, which would result in the request reaching the target user regardless of which device they are currently using, or if they have multiple devices.

SIP uses e-mail-like names for addresses. The addressing scheme is part of a family of Internet addresses known as URIs. SIP URIs can also handle telephone numbers, transport parameters, and a number of other items. A full description, including examples, can be found in Section 4.2. For now, the key point is that a SIP URI is a name that is resolved to an IP address by using SIP proxy server and DNS lookups at the time of the call, as will be seen in the next example. Figure 2.2 shows an example of a more typical SIP call with a type of SIP server called a "proxy server." In this example, the caller Schroedinger calls Heisenberg through a SIP proxy server. A SIP proxy operates in a similar way to a proxy in HTTP and other Internet protocols. A SIP proxy does not set up or terminate sessions, but sits in the middle of a SIP message exchange, receiving messages and forwarding them. This example shows one proxy, but there can be multiple proxies in a signaling path.

SIP has two broad categories of URIs: ones that correspond to a user, and ones that correspond to a single device or end point. The user URI is known as an address of record (AOR) and a request sent to an address of record will require database lookups and service and feature operations and can result the request being sent to one or more end devices. A device URI is known as a contact, and typically does not require database lookups. An address of record URI is usually used in To and From header fields, as this is the general way to reach a person and is suitable for storing in address books and in returning missed calls. A device URI is usually used in a Contact header field and is associated with a particular user for a shorter period of time. The method of relating (or binding) a contact URI with an address of record URI will be discussed in Section 2.3.

Because Schroedinger does not know exactly where Heisenberg is currently logged on and which device they are currently using, a SIP proxy server is used to route the INVITE. First, a DNS lookup of Heisenberg's SIP URI domain name (munich.de) is performed, which returns the IP address of the proxy server proxy.munich.de, which handles that domain. The INVITE is then sent to that IP address:

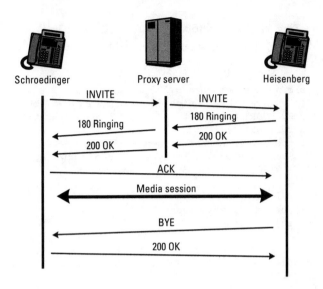

Figure 2.2 SIP call example with proxy server.

```
INVITE sip:werner.heisenberg@munich.de SIP/2.0
Via: SIP/2.0/UDP 100.101.102.103:5060;branch=z9hG4bKmp17a
Max-Forwards: 70
To: Heisenberg <sip:werner.heisenberg@munich.de>
From: E. Schroedinger <sip:schroed5244@aol.com>;tag=42
Call-ID: 10@100.101.102.103
CSeq: 1 INVITE
Subject: Where are you exactly?
Contact: <sip:schroed5244@pc33.aol.com>
Content-Type: application/sdp
Content-Length: 159

v=0
o=schroed5244 2890844526 2890844526 IN IP4 100.101.102.103
s=Phone Call
t=0 0
c=IN IP4 100.101.102.103
m=audio 49170 RTP/AVP 0
a=rtpmap:0 PCMU/8000
```

The proxy looks up the SIP URI in the Request-URI (`sip:wer-ner.heisenberg@munich.de`) in its database and locates Heisenberg. This completes the two-step process:

1. DNS lookup by user agent to locate the IP address of the proxy; database lookup is performed by the proxy to locate the IP address;

2. The INVITE is then forwarded to Heisenberg's IP address with the
    addition of a second Via header field stamped with the address of
    the proxy:

```
INVITE sip:werner.heisenberg@200.201.202.203 SIP/2.0
Via: SIP/2.0/UDP proxy.munich.de:5060;branch=z9hG4bK83842.1
Via: SIP/2.0/UDP 100.101.102.103:5060;branch=z9hG4bKmp17a
Max-Forwards: 69
To: Heisenberg <sip:werner.heisenberg@munich.de>
From: E. Schroedinger <sip:schroed5244@aol.com>;tag=42
Call-ID: 10@100.101.102.103
CSeq: 1 INVITE
Contact: <sip:schroed5244@pc33.aol.com>
Content-Type: application/sdp
Content-Length: 159

v=0
o=schroed5244 2890844526 2890844526 IN IP4 100.101.102.103
s=Phone Call
c=IN IP4 100.101.102.103
t=0 0
m=audio 49172 RTP/AVP 0
a=rtpmap:0 PCMU/8000
```

From the presence of two Via header fields, Heisenberg knows that the
INVITE has been routed through a proxy server. Having received the
INVITE, a 180 Ringing response is sent by Heisenberg to the proxy:

```
SIP/2.0 180 Ringing
Via: SIP/2.0/UDP proxy.munich.de:5060;branch=z9hG4bK83842.1
 ;received=100.101.102.105
Via: SIP/2.0/UDP 100.101.102.103:5060;branch=z9hG4bKmp17a
To: Heisenberg <sip:werner.heisenberg@munich.de>;tag=314159
From: E. Schroedinger <sip:schroed5244@aol.com>;tag=42
Call-ID: 10@100.101.102.103
CSeq: 1 INVITE
Contact: <sip:werner.heisenberg@200.201.202.203>
Content-Length: 0
```

Again, this response contains the Via header fields, and the To,
From, Call-ID, and CSeq header fields from the INVITE request. The
response is then sent to the address in the first Via header field,
proxy.munich.de to the port number listed in the Via header field:
5060, in this case. Notice that the To header field now has a tag added to it to
identify this particular dialog. Only the first Via header field contains a
received parameter, since the second Via header already contains the literal
IP address in the URI. The Contact header field contains the device URI of
Heisenberg.

The proxy receives the response, checks that the first Via header field has
its own address (proxy.munich.de), uses the transaction identifier in the

Via header, then removes that Via header field, then forwards the response to the address in the next Via header field: IP address 100.101.102.103, port 5060. The resulting response sent by the proxy to Schroedinger is:

```
SIP/2.0 180 Ringing
Via: SIP/2.0/UDP 100.101.102.103:5060;branch=z9hG4bKmp17a
To: Heisenberg <sip:werner.heisenberg@munich.de>;tag=314159
From: E. Schroedinger <sip:schroed5244@aol.com>;tag=42
Call-ID: 10@100.101.102.103
CSeq: 1 INVITE
Contact: <sip:werner.heisenberg@200.201.202.203>
Content-Length: 0
```

The use of Via header fields in routing and forwarding SIP messages reduces complexity in message forwarding. The request required a database lookup by the proxy to be routed. The response requires no lookup because the routing is imbedded in the message in the Via header fields. Also, this ensures that responses route back through the same set of proxies as the request. The call is accepted by Heisenberg, who sends a 200 OK response:

```
SIP/2.0 200 OK
Via: SIP/2.0/UDP proxy.munich.de:5060;branch=z9hG4bK83842.1
 ;received=100.101.102.105
Via: SIP/2.0/UDP 100.101.102.103:5060;branch=z9hG4bKmp17a
To: Heisenberg <sip:werner.heisenberg@munich.de>;tag=314159
From: E. Schroedinger <sip:schroed5244@aol.com>;tag=42
Call-ID: 10@100.101.102.103
CSeq: 1 INVITE
Contact: <sip:werner.heisenberg@200.201.202.203>
Content-Type: application/sdp
Content-Length: 159

v=0
o=heisenberg 2890844526 2890844526 IN IP4 200.201.202.203
s=Phone Call
c=IN IP4 200.201.202.203
t=0 0
m=audio 49172 RTP/AVP 0
a=rtpmap:0 PCMU/8000
```

The proxy forwards the 200 OK message to Schroedinger after removing the first Via header field:

```
SIP/2.0 200 OK
Via: SIP/2.0/UDP 100.101.102.103:5060;branch=z9hG4bKmp17a
To: Heisenberg <sip:werner.heisenberg@munich.de>;tag=314159
From: E. Schroedinger <sip:schroed5244@aol.com>;tag=42
Call-ID: 10@100.101.102.103
CSeq: 1 INVITE
Contact: <sip:werner.heisenberg@200.201.202.203>
Content-Type: application/sdp
```

```
Content-Length: 159

v=0
o=heisenberg 2890844526 2890844526 IN IP4 200.201.202.203
c=IN IP4 200.201.202.203
t=0 0
m=audio 49170 RTP/AVP 0
a=rtpmap:0 PCMU/8000
```

The presence of the Contact header field with the SIP URI address of Heisenberg in the 200 OK allows Schroedinger to send the ACK directly to Heisenberg, bypassing the proxy. (Note that the Request-URI is set to Heisenberg's Contact URI and not the URI in to To header field.) This request, and all future requests continue to use the tag in the To header field:

```
ACK sip:werner.heisenberg@200.201.202.203 SIP/2.0
Via: SIP/2.0/UDP 100.101.102.103:5060;branch=z9hG4bKKka42
Max-Forwards: 70
To: Heisenberg <sip:werner.heisenberg@munich.de>;tag=314159
From: E. Schroedinger <sip:schroed5244@aol.com>;tag=42
Call-ID: 10@100.101.102.103
CSeq: 1 ACK
Content-Length: 0
```

This shows that the proxy server is not really "in the call." It facilitates the two end points locating and contacting each other, but it can drop out of the signaling path as soon as it no longer adds any value to the exchange. A proxy server can force further messaging to route through it by inserting a Record-Route header field, which is described in Section 6.2.12. In addition, it is possible to have a proxy server that does not retain any knowledge of the fact that there is a session established between Schroedinger and Heisenberg (referred to as *call state information*). This is discussed in Section 3.3.1. Note that the media is always end-to-end and not through the proxy.

In SIP the path of the signaling messages is totally independent of the path of the media. In telephony, this is described as the separation of control channel and bearer channel.

The media session is ended when Heisenberg sends a BYE message:

```
BYE sip:schroed5244@pc33.aol.com SIP/2.0
Via: SIP/2.0/UDP 200.201.202.203:5060;branch=z9hG4bK4332
Max-Forwards: 70
To: E. Schroedinger <sip:schroed5244@aol.com>;tag=42
From: Heisenberg <sip:werner.heisenberg@munich.de>;tag=314159
Call-ID: 10@100.101.102.103
CSeq: 2000 BYE
Content-Length: 0
```

Note that Heisenberg's CSeq was initialized to 2000. Each SIP device maintains its own independent CSeq number space. This is explained in some detail in Section 6.1.5. The Request-URI is set to Schroedinger's Contact URI. Schroedinger confirms with a 200 OK response:

```
SIP/2.0 200 OK
Via: SIP/2.0/UDP 200.201.202.203:5060;branch=z9hG4bK4332
To: E. Schroedinger <sip:schroed5244@aol.com>;tag=42
From: Heisenberg <sip:werner.heisenberg@munich.de>;tag=314159
Call-ID: 10@100.101.102.103
CSeq: 2000 BYE
Content-Length: 0
```

Not discussed in the previous example is the question of how the database accessed by the proxy contained Heisenberg's current IP address. There are many ways this could be done using SIP or other protocols. The mechanism for accomplishing this using SIP is called *registration* and is discussed in the next section.

2.3 SIP Registration Example

In this example, shown in Figure 2.3, Heisenberg sends a SIP REGISTER request to a type of SIP server known as a registrar server. The SIP registrar server receives the message uses the information in the request to update the database used by proxies to route SIP requests. Contained in the REGISTER message To header is the SIP URI address of Heisenberg. This is Heisenberg's "well-known" address, perhaps printed on his business card or published on a Web page or in a directory. Also contained in the REGISTER is a Contact URI, which represents the current device (and its IP address) that the user

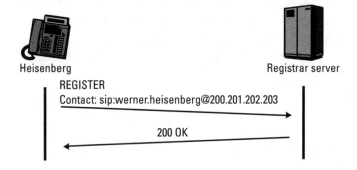

Figure 2.3 SIP registration example.

Heisenberg is currently using. The registrar binds the SIP URI of Heisenberg and the IP address of the device in a database that can be used, for example, by the proxy server in Figure 2.2 to locate Heisenberg. When a proxy server with access to the database receives an INVITE request addressed to Heisenberg's URI (i.e., an incoming call), the request will be proxied to the Contact URI of the currently registered device.

This registration has no real counterpart in the telephone network, but it is very similar to the registration a wireless phone performs when it is turned on. A cell phone sends its identity to the base station (BS), which then forwards the location and phone number of the cell phone to a home location register (HLR). When the mobile switching center (MSC) receives an incoming call, it consults the HLR to get the current location of the cell phone. Further aspects of SIP mobility are discussed in Chapter 9.

The REGISTER message sent by Heisenberg to the SIP registrar server has the form:

```
REGISTER sip:registrar.munich.de SIP/2.0
Via: SIP/2.0/UDP 200.201.202.203:5060;branch=z9hG4bKus19
Max-Forwards: 70
To: Werner Heisenberg <sip:werner.heisenberg@munich.de>
From: Werner Heisenberg <sip:werner.heisenberg@munich.de>
 ;tag=3431
Call-ID: 23@200.201.202.203
CSeq: 1 REGISTER
Contact: sip:werner.heisenberg@200.201.202.203
Content-Length: 0
```

The Request-URI in the start line of the message contains the address of the registrar server. In a REGISTER request, the To header field contains the URI that is being registered, in this case sip:werner.heisenberg@munich.de. This results in the To and From header fields usually being the same, although an example of third-party registration is given in Section 4.1.2. The SIP URI in the Contact address is stored by the registrar.

The registrar server acknowledges the successful registration by sending a 200 OK response to Heisenberg. The response echoes the Contact information that has just been stored in the database and includes a To tag:

```
SIP/2.0 200 OK
Via: SIP/2.0/UDP 200.201.202.203:5060;branch=z9hG4bKus19
To: Werner Heisenberg <sip:werner.heisenberg@munich.de>;tag=8771
From: Werner Heisenberg <sip:werner.heisenberg@munich.de>
 ;tag=3431
Call-ID: 23@200.201.202.203
CSeq: 1 REGISTER
Contact: <sip:werner.heisenberg@munich.de>;expires=3600
Content-Length: 0
```

The `Contact` URI is returned along with an `expires` parameter, which indicates how long the registration is valid, which in this case is 1 hour (3,600 seconds). If Heisenberg wants the registration to be valid beyond that interval, he must send another `REGISTER` request within the expiration interval.

Registration is typically automatically performed on initialization of a SIP device and at regular intervals determined by the expiration interval chosen by the registrar server. Registration is an additive process—more than one device can be registered against a SIP URI. If more than one device is registered, a proxy may forward the request to either or both devices, either in a sequential or parallel search. Additional register operations can be used to clear registrations or retrieve a list of currently registered devices.

2.4 SIP Presence and Instant Message Example

This example shows how SIP is used in a presence and instant messaging application. Presence information can be thought of as the state of a user or device at a particular instant. It can be as simple as whether a particular user is signed in or not, whether they are active at their station, or idle or away. For a mobile device, presence information can include the actual location in terms of coordinates, or in general terms such as "in the office," "on travel," or "in the lab." Presence information can even include information about the status or mood of the user, whether they are working, relaxing, or socializing. For all these examples, a presence protocol is mainly concerned about establishing subscriptions or long-term relationships between devices about transferring status information, and the delivery of that information. The actual information transferred, and how that information is presented to the user is application dependent. In terms of the SIP protocol, `SUBSCRIBE` is used to request status or presence updates from the presence server (or "presentity"), and `NOTIFY` is used to deliver that information to the requestor or presence "watcher." SIP presence uses the SIP Events extensions [4] and instant message extensions [5].

In this example, Chebychev wishes to communicate with Poisson. The message flow is shown in Figure 2.4. To find out the status of Poisson, Chebychev subscribes to Poisson's presence information by sending a `SUBSCRIBE` message to Poisson. The request looks like:

```
SUBSCRIBE sip:poisson@probability.org SIP/2.0
Via SIP/2.0/TCP lecturehall21.academy.ru:5060
  ;branch=z9hG4bK348471123
Max-Forwards: 70
To: M. Poisson <sip:poisson@probability.org>
From: P. L. Chebychev <sip:chebychev@academy.ru>;tag=21171
Call-ID: 58dkfj349241k34452k592520
CSeq: 3412 SUBSCRIBE
Allow-Events: presence
```

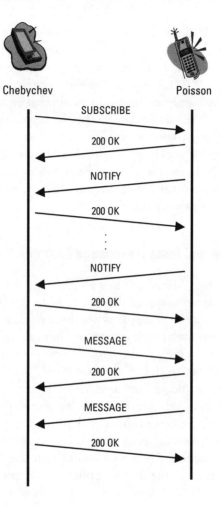

Figure 2.4 SIP presence and instant message example.

```
Allow: ACK, INVITE, CANCEL, BYE, NOTIFY, SUBSCRIBE, MESSAGE
Contact: <sip:pafnuty@lecturehall21.academy.ru;transport=tcp>
Event: presence
Content-Length: 0
```

In this example, TCP is used as the transport for the SIP messages as indicated in the Via header field and in the transport=tcp parameter in the Contact URI. This request also contains Allow and Allow-Events header fields, which are used to advertise capabilities. In this example, Chebychev is indicating support for receiving seven methods listed in the Allow header field, and also presence subscriptions in the Allow-Event header field. As this SUBSCRIBE is creating a dialog (in an analogous way that an

`INVITE` created a dialog in the earlier examples), the `From` contains a tag but the `To` header field does not yet contain a tag.

Poisson accepts the subscription request by sending a `202 Accepted` response back to Chebychev:

```
SIP/2.0 202 Accepted
Via SIP/2.0/TCP lecturehall21.academy.ru:5060
 ;branch=z9hG4bK348471123;received=19.34.3.1
To: M. Poisson <sip:poisson@probability.org>;tag=25140
From: P. L. Chebychev <sip:chebychev@academy.ru>;tag=21171
Call-ID: 58dkfj349241k34452k592520
CSeq: 3412 SUBSCRIBE
Allow-Events: presence
Allow: ACK, INVITE, CANCEL, BYE, NOTIFY, SUBSCRIBE, MESSAGE
Contact: <sip:s.possion@dist.probability.org;transport=tcp>
Event: presence
Expires: 3600
Content-Length: 0
```

In this example, there are no proxy servers between Chebychev's watcher and Poisson's presence server, although there could be any number. The `Expires` header field indicates that the subscription expires in 1 hour. The actual subscription is begun by Poisson sending the first `NOTIFY` back to Chebychev:

```
NOTIFY sip:pafnuty@lecturehall21.academy.ru SIP/2.0
Via SIP/2.0/TCP dist.probablilty.org:5060
 ;branch=z9hG4bK4321
Max-Forwards: 70
To: P. L. Chebychev <sip:chebychev@academy.ru>;tag=21171
From: M. Poisson <sip:poisson@probability.org>;tag=25140
Call-ID: 58dkfj349241k34452k592520
CSeq: 1026 NOTIFY
Allow: ACK, INVITE, CANCEL, BYE, NOTIFY, SUBSCRIBE, MESSAGE
Allow-Events: dialog
Contact: <sip:s.possion@dist.probability.org;transport=tcp>
Subscription-State: active;expires=3600
Event: presence
Content-Type: application/cpim-pidf+xml
Content-Length: 244

<?xml version="1.0" encoding="UTF-8"?>
<presence xmlns="urn:ietf:params:xml:ns:cpim-pidf"
  entity="sip:poisson@probability.org">
 <tuple id="452426775">
  <status>
   <basic>closed</basic>
  </status>
 </tuple>
</presence>
```

Note that this NOTIFY is sent within the dialog established with the SUBSCRIBE—it uses the same dialog identifier (Call-ID, local and remote tags)—and the request is sent to the Contact URI provided by Chebychev in the subscription request. The Subscription-State header field indicates that the subscription has been authorized and activate and that it will expire in 1 hour unless refreshed by Chebychev (using another SUBSCRIBE request).

The Common Presence and Instant Message Presence Information Data Format (CPIM PIDF) [6] XML message body contains the status information that Poisson is currently off-line (closed).

Chebychev sends a 200 OK response to the NOTIFY to confirm that it has been received:

```
SIP/2.0 200 OK
Via SIP/2.0/TCP dist.probablilty.org:5060
 ;branch=z9hG4bK4321;received=24.32.1.3
To: P. L. Chebychev <sip:chebychev@academy.ru>;tag=21171
From: M. Poisson <sip:poisson@probability.org>;tag=25140
Call-ID: 58dkfj349241k34452k592520
CSeq: 1026 NOTIFY
Content-Length: 0
```

Later, when Poisson does sign in, this information is provided in a second NOTIFY containing the change in status:

```
NOTIFY sip:pafnuty@lecturehall21.academy.ru SIP/2.0
Via SIP/2.0/TCP dist.probablilty.org:5060
 ;branch=z9hG4bK334241
Max-Forwards: 70
To: P. L. Chebychev <sip:chebychev@academy.ru>;tag=21171
From: M. Poisson <sip:poisson@probability.org>;tag=25140
Call-ID: 58dkfj349241k34452k592520
CSeq: 1027 NOTIFY
Allow: ACK, INVITE, CANCEL, BYE, NOTIFY, SUBSCRIBE, MESSAGE
Allow-Events: presence
Contact: <sip:s.possion@dist.probability.org;transport=tcp>
Subscription-State: active;expires=1800
Event: presence
Content-Type: application/cpim-pidf+xml
Content-Length: 325

<?xml version="1.0" encoding="UTF-8"?>
<presence xmlns="urn:ietf:params:xml:ns:cpim-pidf"
  entity="sip:poisson@probability.org">
 <tuple id="452426775">
  <status>
   <basic>open</basic>
  </status>
   <contact>sip:s.possion@dist.probability.org;transport=tcp
   </contact>
```

```
    </tuple>
  </presence>
```

The expiration time indicated in the `Subscription-State` header field indicates that 30 minutes have passed since the subscription was established. The CPIM PIDF XML message body now indicates that Poisson is on-line (open) and can be reached via the URI `sip:s.possion@dist.probability.org;transport=tcp`.

Chebychev confirms receipt of the `NOTIFY` with a `200 OK` response:

```
SIP/2.0 200 OK
Via SIP/2.0/TCP dist.probablilty.org:5060
 ;branch=z9hG4bK334241;received=24.32.1.3
To: P. L. Chebychev <sip:chebychev@academy.ru>;tag=21171
From: M. Poisson <sip:poisson@probability.org>;tag=25140
Call-ID: 58dkfj349241k34452k592520
CSeq: 1027 NOTIFY
Content-Length: 0
```

Now that Chebychev knows that Poisson is on-line, he sends an instant message to him using the `Contact` URI from the `NOTIFY`:

```
MESSAGE sip:s.possion@dist.probability.org SIP/2.0
Via SIP/2.0/TCP lecturehall21.academy.ru:5060
 ;branch=z9hG4bK3gtr2
Max-Forwards: 70
To: M. Poisson <sip:s.possion@dist.probability.org>
From: P. L. Chebychev <sip:chebychev@academy.ru>;tag=4542
Call-ID: 9dkei93vjq1ei3
CSeq: 15 MESSAGE
Allow: ACK, INVITE, CANCEL, BYE, NOTIFY, SUBSCRIBE, MESSAGE
Content-Type: text/plain
Content-Length: 9

Hi There!
```

Notice that this `MESSAGE` is sent outside the dialog. Instant messages sent using the `MESSAGE` method in SIP are like page messages—they are not part of any dialog. As a result, each message contains a new `Call-ID` and `From` tag. The `200 OK` response is used to acknowledge receipt of the instant message:

```
SIP/2.0 200 OK
Via SIP/2.0/TCP lecturehall21.academy.ru:5060
 ;branch=z9hG4bK3gtr2;received=19.34.3.1
To: M. Poisson <sip:s.possion@dist.probability.org>;tag=2321
From: P. L. Chebychev <sip:chebychev@academy.ru>;tag=4542
Call-ID: 9dkei93vjq1ei3
CSeq: 15 MESSAGE
Content-Length: 0
```

Poison answers with a reply, which is also sent outside of any dialog, with a new `Call-ID` and `From` tag (an instant message response is never sent in a `200 OK` reply to a `MESSAGE` request):

```
MESSAGE sip:chebychev@academy.ru SIP/2.0
Via SIP/2.0/TCP dist.probablilty.org:5060
 ;branch=z9hG4bK4526245
Max-Forwards: 70
To: P. L. Chebychev <sip:chebychev@academy.ru>
From: M. Poisson <sip:s.possion@dist.probability.org>;tag=14083
Call-ID: 1k34452k592520
CSeq: 2321 MESSAGE
Allow: ACK, INVITE, CANCEL, BYE, NOTIFY, SUBSCRIBE, MESSAGE
Content-Type: text/plain
Content-Length: 30

Well, hello there to you, too!
```

which receives a `200 OK` reply:

```
SIP/2.0 200 OK
Via SIP/2.0/TCP dist.probablilty.org:5060
 ;branch=z9hG4bK4526245;received=24.32.1.3
To: P. L. Chebychev <sip:chebychev@academy.ru>;tag=mc3bg5q77wms
From: M. Poisson <sip:s.possion@dist.probability.org>;tag=14083
Call-ID: 1k34452k592520
CSeq: 2321 MESSAGE
Content-Length: 0
```

Other presence packages define other sets of information that can be requested by watchers from presence servers.

2.5 Message Transport

As discussed in Chapter 1, SIP is a layer four, or application layer, protocol in the Internet Multimedia Protocol stack shown in Figure 1.1. RFC 3261 defines the use of TCP, UDP, or TLS transport. An extension document describes how SCTP can be used. How a SIP message is transported using these four protocols will be described in the following sections. The compression of SIP for transport over low bandwidth connections, such as wireless, is discussed in Chapter 9.

2.5.1 UDP Transport

When using UDP, each SIP request or response message is carried in a single UDP datagram or packet. For a particularly large message body, there is a compact form of SIP that saves space in representing some header fields with a single character. This is discussed in Chapter 6. Figure 2.5 shows a SIP BYE request exchange during an established SIP session using UDP.

The source port is chosen from a pool of available port numbers (above 49172), or sometimes the default SIP port of 5060 is used. The lack of hand-shaking or acknowledgment in UDP transport means that a datagram could be lost and a SIP message along with it. The checksum, however, enables UDP to discard errored datagrams, allowing SIP to assume that a received message is complete and error-free. The reliability mechanisms built into SIP to handle message retransmissions are described in Section 3.5. The reply is also sent to port 5060, or the port number listed in the top Via header field.

UDP provides the simplest transport for user agents and servers and allows them to operate without transport layer state. However, UDP offers no congestion control. A series of lost packets on a heavily loaded IP link can cause retransmissions, which in turn produce more lost packets and can push the link into congestion collapse. Also, UDP may only be used for SIP when the message (and its response) is known to be less than the Message Transport Unit (MTU) size of the IP network. For simple SIP messages, this is not a problem. However, for large messages containing multiple message bodies and large header fields, this can be a problem. In this case, TCP must be used, since SIP does not support fragmentation at the SIP layer.

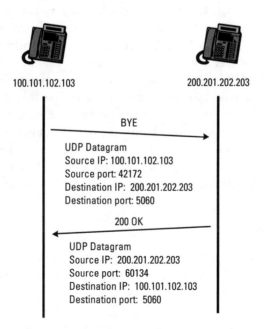

Figure 2.5 Transmission of SIP messages using UDP.

2.5.2   TCP Transport

TCP provides a reliable transport layer, but at a cost of complexity and transmission delay over the network. The use of TCP for transport in a SIP message exchange is shown in Figure 2.6. This example shows an INVITE sent by a user agent at 100.101.102.103 to a type of SIP server called a redirect server at 200.201.202.203. A SIP redirect server does not forward INVITE requests like a proxy, but looks up the destination address and instead returns that address in a redirection class (3xx) response. The 302 Moved Temporarily response is acknowledged by the user agent with an ACK message. Not shown in this figure is the next step, where the INVITE would be re-sent to the address returned by the redirect server. As in the UDP example, the well-known SIP port number of 5060 is chosen for the destination port, and the source port is chosen from an available pool of port numbers. Before the message can be sent, however, the TCP connection must be opened between the two end points. This transport layer datagram exchange is shown in Figure 2.6 as a single arrow, but it is actually a three-way handshake between the end points as shown in Figure 1.2. Once the connection is established, the messages are sent in the stream.

The Content-Length header field is critical when TCP is used to transport SIP, since it is used to find the end of one message and the start of the next. When TCP or another stream-based transport is used, Content-Length is a required header field in all requests and responses.

To send the 302 Moved Temporarily response, the server typically opens a new TCP connection in the reverse direction, using 5060 (or the port listed in the top Via header field) as the destination port.[2] The acknowledgment ACK is sent in the TCP stream used for the INVITE. Because this concludes the SIP session, the connection is then closed. If a TCP connection closes during a dialog, a new one can be opened to send a request within the dialog, such as a BYE request to terminate the media session.

As previously mentioned, TCP provides reliable transport and also congestion control. It can also transport SIP messages of arbitrary size. The disadvantages of TCP include the setup delay in establishing the connection and the need for servers to maintain this connection state at the transport layer.

2.5.3   TLS Transport

New to RFC 3261 is support of TLS version 1.0 [8] in the SIP specification. SIP can use TLS over TCP as for encrypted transport with the additional capabilities

---

2. There is work underway to enable TCP connection reuse [7]. This would allow responses to be sent over the same connection as the request, and also allow multiple transactions and dialogs to share a TCP connection, for example, between two proxy servers.

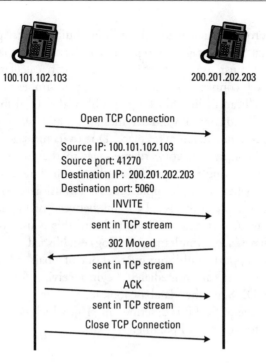

100.101.102.103   200.201.202.203

Open TCP Connection

Source IP: 100.101.102.103
Source port: 41270
Destination IP: 200.201.202.203
Destination port: 5060

INVITE

sent in TCP stream

302 Moved

sent in TCP stream

ACK

sent in TCP stream

Close TCP Connection

Figure 2.6 Transmission of SIP messages using TCP.

of authentication. In Section 4.2.1 the secure SIP URI scheme (`sips`) will be discussed, which uses TLS transport. The default SIP port number for TLS transport is port 5061.

If TLS is used between two proxies, each proxy may have a certificate allowing mutual authentication. However, if a client does not have a certificate, TLS can be used in conjunction with another authentication mechanism, such as SIP digest, to allow mutual authentication.

The SIP use of TLS takes advantage of both the encryption and authentication services. However, the encryption and authentication is only useful on a single hop. If a SIP request takes multiple hops (i.e., includes one or more proxy servers), TLS is not useful for end-to-end authentication. S/MIME encryption, described in Section 3.7 solves this problem.

SIP Proxies must support TLS and will likely use TLS for long-lived connections.

2.5.4 SCTP Transport

An extension to SIP defines the use of SCTP [9] with SIP to provide reliable stream-based transport with some advantages over TCP transport for a message-based protocol such as SIP. First, it has built-in message segmentation, so that individual SIP messages are separated at the transport layer. With TCP, the SIP protocol must use the `Content-Length` calculation to delineate messages. If a TCP connection is being shared by a number of SIP transactions and dialogs, the "head of line blocking" problem discussed in Section 1.4.3.4 can cause the buffer to contain valid SIP messages that could be processed by the server while the retransmission takes place. Due to its message level delineation, SCTP is able to continue to forward messages to the application layer while simultaneously requesting a retransmission of a dropped message. Note that this is only a problem when multiple applications are multiplexed over a single TCP connection. An example of this is a TCP link between two signaling proxy servers. For a user agent to proxy TCP connection, this is usually not a problem unless the two have many simultaneous dialogs established.

SCTP also supports multihoming, so if one of a pair of load balancing SIP proxies fails, the other can immediately begin receiving the messages without even requiring a DNS or other database lookup.

The SIP usage of SCTP is described in [10], which defines the syntax for the `transport=sctp` URI parameter.

References

[1]   Rosenberg, J., et al., "SIP: Session Initiation Protocol," RFC 3261, 2002.

[2]   Schulzrinne, H., et al., "RTP: A Transport Protocol for Real-Time Applications," RFC 3550, 2003.

[3]   Handley, M., and V. Jacobson, "SDP: Session Description Protocol," RFC 2327, 1998.

[4]   Roach, A., "Session Initiation Protocol (SIP)-Specific Event Notification," RFC 3265, 2002.

[5]   Campbell, B., et al., "Session Initiation Protocol (SIP) Extension for Instant Messaging," RFC 3428, 2002.

[6]   Sugano, H., et al., "Common Presence and Instant Messaging (CPIM) Presence Information Data Format," IETF Internet-Draft, Work in Progress, 2002.

[7]   Mahy, R., "Requirements for Connection Reuse in the Session Initiation Protocol (SIP)," IETF Internet-Draft, Work in Progress, 2003

[8]   Dierks, T., et al., "The TLS Protocol Version 1.0," RFC 2246, 1999.

[9]   Stewart, R., et al., "Stream Control Transmission Protocol," RFC 2960, 1999.

[10]   Rosenberg, J., H. Schulzrinne, and G. Camarillo, "The Stream Control Transmission Protocol as a Transport for the Session Initiation Protocol," IETF Internet-Draft, Work in Progress, 2002.

3

SIP Clients and Servers

The client-server nature of SIP has been introduced in the example message flows of Chapter 2. In this chapter, the types of clients and servers in a SIP network will be introduced and defined.

3.1   SIP User Agents

A SIP-enabled end-device is called a SIP user agent [1]. One purpose of SIP is to enable sessions to be established between user agents. As the name implies, a user agent takes direction or input from a user and acts as an agent on their behalf to set up and tear down media sessions with other user agents. In most cases, the user will be a human, but the user could be another protocol, as in the case of a gateway (described in the next section). A user agent must be capable of establishing a media session with another user agent.

A UA must maintain state on calls that it initiates or participates in. A minimum call state set includes the local and remote tags, Call-ID, local and remote CSeq header fields, along with the route set and any state information necessary for the media. This information is used to store the dialog information and for reliability. The remote CSeq storage is necessary to distinguish between a re-INVITE and a retransmission. A re-INVITE is used to change the session parameters of an existing or pending call. It uses the same Call-ID, but the CSeq is incremented because it is a new request. A retransmitted INVITE will contain the same Call-ID and CSeq as a previous INVITE. Even after a call has been terminated, call state must be maintained by a user agent for at least 32 seconds in case of lost messages in the call tear-down.

User agents silently discard an ACK for an unknown dialog. Requests to an unknown URI receive a 404 Not Found Response. A user agent receiving a request for an unknown dialog responds with a 481 Dialog/Transaction Does Not Exist. Responses from an unknown dialog are also silently discarded. These silent discards are necessary for security. Otherwise, a malicious user agent could gain information about other SIP user agents by spamming fake requests or responses.

Although not required to understand every response code defined, a minimal implementation must to be able to interpret any unknown response based on the class (first digit of the number) of the response. That is, if an undefined 498 Wrong Phase of the Moon response is received, it must be treated as a 400 Client Error.

A user agent responds to an unsupported request with a 501 Not Implemented response. A SIP UA must support UDP transport and also TCP if it sends messages greater than 1,000 octets in size.

A SIP user agent contains both a client application and a server application. The two parts are a user agent client (UAC) and user agent server (UAS). The UAC initiates requests while the UAS generates responses. During a session, a user agent will usually operate as both a UAC and a UAS.

A SIP user agent must also support SDP for media description. Other types of media description protocols can be used in bodies, but SDP support is mandatory. Details of SDP are in Section 7.1.

A UA must understand any extensions listed in a Require header field in a request. Unknown header fields may be ignored by a UA.

A UA should advertise its capabilities and features in any request it sends. This allows other UAs to learn of them without having to make an explicit capabilities query. For example, the methods that a UA supports should be listed in an Allow header field. SIP extensions should be listed in a Supported header field. Message body types that are supported should be listed in an Accept header field.

3.2 Presence Agents

A presence agent (PA) [2] is a SIP device that is capable of receiving subscription requests and generating state notifications as defined by the SIP Events specification [3]. An example of a presence agent is in the example of Section 2.4. A presence agent supports the presence Event package [2], responds to SUBSCRIBE requests, and sends NOTIFY requests.

A presence agent can collect presence information from a number of devices. Presence information can come from a SIP device registering, a SIP device publishing presence information [4], or from many other non-SIP sources.

A presence server is a server that sometimes acts as a presence agent and supplies presence information and other times acts as a proxy, forwarding SUBSCRIBE requests to another presence agent.

A presence agent first authenticates a subscription request. If the authentication passes, it establishes a dialog and sends the notifications over that dialog. The subscription can be refreshed by receiving new SUBSCRIBE requests.

3.3 Back-to-Back User Agents

A back-to-back user agent (B2BUA) is a type of SIP device that receives a SIP request, then reformulates the request and sends it out as a new request. Responses to the request are also reformulated and sent back in the opposite direction. For example, a B2BUA device can be used to implement an anonymizer service in which two SIP UAs can communicate without either party learning the other party's URI, IP address, or any other information. To achieve this, an anonymizer B2BUA would reformulate a request with an entirely new From, Via, Contact, Call-ID, and SDP media information, also removing any other SIP header fields that might contain information about the calling party. The response returned would also change the Contact and SDP media information from the called party. The modified SDP would point to the B2BUA itself, which would forward RTP media packets from the called party to the calling party and vice-versa. In this way, neither end point learns any identifying information about the other party during the session establishment. (Of course, the calling party needs to know the called party's URI in order for the call to take place.)

Sometimes B2BUAs are employed to implement other SIP services. However, they break the end-to-end nature of an Internet protocol such as SIP. Also, a B2BUA is a call-stateful single point of failure in a network, which means their use will reduce the reliability of SIP sessions over the Internet. The relayed media suffers from increased latency and increased probability of packet loss, which can reduce the quality of the media session. Geographic distribution of B2BUAs can reduce these effects, but the problem of selecting the best B2BUA for a particular session is a very difficult one since the source and destination IP address of the media are not known until the session is actually established (with a 200 OK).

The most common form of B2BUA present in SIP networks is application layer gateways (ALG). Some firewalls have ALG functionality built in, which allows a firewall to permit SIP and media traffic while still maintaining a high level of security. The use of ALGs is described in Section 3.11.

3.4 SIP Gateways

A SIP gateway is an application that interfaces a SIP network to a network utilizing another signaling protocol. In terms of the SIP protocol, a gateway is just a special type of user agent, where the user agent acts on behalf of another protocol rather than a human. A gateway terminates the signaling path and can also terminate the media path, although this is not always the case. For example, a SIP to H.323 gateway terminates the SIP signaling path and converts the signaling to H.323, but the SIP user agent and H.323 terminal can exchange RTP media information directly with each other without going through the gateway. An example of this is described in Section 10.6.

A SIP to Public Switched Telephone Network (PSTN) gateway terminates both the signaling and media paths. SIP can be translated into, or interwork with, common PSTN protocols such as Integrated Services Digital Network (ISDN), ISDN User Part (ISUP), and other Circuit Associated Signaling (CAS) protocols, which are briefly described in Section 7.4. A PSTN gateway also converts the RTP media stream in the IP network into a standard telephony trunk or line. The conversion of signaling and media paths allows calling to and from the PSTN using SIP. Examples of these gateways are described in Sections 10.3 and 10.4. Figure 3.1 shows a SIP network connected via gateways with the PSTN and a H.323 network. There is work underway to standardize the SIP/H.323 interworking function [5].

In the figure, the SIP network, PSTN network, and H.323 networks are shown as clouds, which obscure the underlying details. Shown connecting to the SIP cloud are SIP IP telephones, SIP-enabled PCs, and corporate SIP gateways with attached telephones. The clouds are connected by gateways. Shown

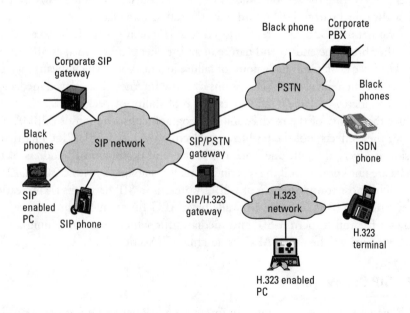

Figure 3.1 SIP network with gateways.

attached to the H.323 network are H.323 terminals and H.323-enabled PCs. The PSTN cloud connects to ordinary analog black telephones (so-called because of the original color of their shell), digital ISDN telephones, and corporate private branch exchanges (PBXs). PBXs connect to the PSTN using shared trunks and provide line interfaces for either analog or digital telephones.

Gateways are sometimes decomposed into a media gateway (MG) and a media gateway controller (MGC). An MGC is sometimes called a call agent because it manages call control protocols (signaling), while the MG manages the media connection. This decomposition is transparent to SIP, and the protocols used to decompose a gateway are not described in this book.

Another difference between a user agent and a gateway is the number of users supported. While a user agent typically supports a single user (although perhaps with multiple lines), a gateway can support hundreds or thousands of users. A PSTN gateway could support a large corporate customer, or an entire geographic area. As a result, a gateway does not REGISTER every user it supports in the same way that a user agent might. Instead, a non-SIP protocol can be used to inform proxies about gateways and assist in routing. One protocol that has been proposed for this is the Telephony Routing over IP (TRIP) protocol [6], which allows an interdomain routing table of gateways to be developed. Another protocol called Telephony Gateway Registration Protocol (TGREP) [7] has also been developed to allow a gateway to register with a proxy server within a domain.

3.5 SIP Servers

SIP servers are applications that accept SIP requests and respond to them. A SIP server should not be confused with a user agent server or the client-server nature of the protocol, which describe operation in terms of clients (originators of requests) and servers (originators of responses to requests). A SIP server is a different type of entity. The types of SIP servers discussed in this section are logical entities. Actual SIP server implementations may contain a number of server types, or may operate as a different type of server under different conditions. Because servers provide services and features to user agents, they must support both TCP, TLS, and UDP for transport. Figure 3.2 shows the interaction of user agents, servers, and a location service. Note that the protocol used between a server and the location service or database is not in general SIP and is not discussed in this book.

3.5.1 Proxy Servers

A SIP proxy server receives a SIP request from a user agent or another proxy and acts on behalf of the user agent in forwarding or responding to the request. A

representing the initial maximum number of unacknowledged segments is sent, which starts at 1 and increases exponentially up to a maximum limit. When network congestion and packet loss occur, the window resets back to 1 and gradually ramps back up again to the maximum limit. A TCP segment header contains 24 octets. Errored segments are detected by a checksum covering both the TCP header and payload.

1.4.3.2 UDP

UDP [10] provides unreliable transport across the Internet. It is a best-effort delivery service, since there is no acknowledgment of sent datagrams. Most of the complexity of TCP is not present, including sequence numbers, acknowledgments, and window sizes. UDP does detect errored datagrams with a checksum. It is up to higher layer protocols, however, to detect this datagram loss and initiate a retransmission if desired.

1.4.3.3 TLS

TLS [11] is based on the Secure Sockets Layer (SSL) protocol first used in Web browsers, and it uses TCP for transport. TLS is commonly used today on the Internet for secure Web sites using the secure HTTP (https) URI scheme.

The TLS protocol has two layers: the TLS Transport protocol and the TLS Handshake protocol. The TLS Transport protocol is used to provide a reliable and private transport mechanism. Data sent using the TLS Transport protocol is encrypted so that a third party cannot intercept the data. A third party also cannot modify the transported data without one of the parties discovering this. The TLS Handshake protocol is used to establish the connection, negotiate the encryption keys used by the TLS Transport protocol, and provide authentication.

The key negotiation scheme selects an encryption algorithm and generates a one-time key based on a secret passed between the two sides. During the handshake, the parties exchange certificates, which can be used for authentication.

The cryptographic computations for a TLS connection are not trivial, and the multiple round trips needed to open a connection as shown in Figure 1.3 can add to message latency. Also, certificate verification can introduce processing delays. However, TLS transport has clear security advantages over UDP or TCP. Many operating systems already support TLS due to its wide use in secure Web browsers and servers.

1.4.3.4 SCTP

The Stream Control Transport Protocol (SCTP) [12] is similar to TCP in that it provides reliable stream-based transport. However, it has some advantages over TCP transport for a message-based protocol. First, it has built-in message segmentation, so that individual messages are separated at the transport layer.

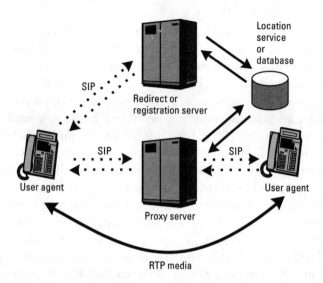

Figure 3.2 SIP user agent, server, and location service interaction.

proxy is not a B2BUA since it is only allowed to modify requests and responses according to strict rules set out in RFC 3261. These rules preserve the end-to-end transparency of the SIP signaling while still allowing a proxy server to perform valuable services and functions for user agents.

A proxy server typically has access to a database or a location service to aid it in processing the request (determining the next hop). The interface between the proxy and the location service is not defined by the SIP protocol. A proxy can use any number of types of databases to aid in processing a request. Databases could contain SIP registrations, presence information, or any other type of information about where a user is located. The example of Figure 2.2 introduced a proxy server as a facilitator of SIP message exchange providing user location services to the caller.

A proxy does not need to understand a SIP request in order to forward it—any unknown request type is assumed to use the non-INVITE transaction model. A proxy should not change the order of header fields or in general modify or delete header fields.

A proxy server is different from a user agent or gateway in three key ways:

1. A proxy server does not issue requests; it only responds to requests from a user agent. (A CANCEL request is an exception to this rule.)

2. A proxy server has no media capabilities.

3. A proxy server does not parse message bodies; it relies exclusively on header fields.

Figure 3.3 shows a common network topology known as the SIP Trapezoid. In this topology, a pair of user agents in different domains establishes a session using a pair of proxy servers, one in each domain. The trapezoid refers to the shape formed by the signaling and media messages. In this configuration, each user agent is configured with a default outbound proxy server, to which it sends all requests. This proxy server typically will authenticate the user agent and may pull up a profile of the user and apply outbound routing services. In an interdomain exchange, DNS SRV queries will be used to locate a proxy server in the other domain. This proxy, sometimes called an inbound proxy may apply inbound routing services on behalf of the called party. This proxy also has access to the current registration information for the user, and can route the request to the called party. In general, future SIP requests will be sent directly between the two user agents, unless one or both proxies inserts a Record-Route header field.

A proxy server can be either stateless or stateful. A stateless proxy server processes each SIP request or response based solely on the message contents. Once the message has been parsed, processed, and forwarded or responded to, no information about the message is stored—no dialog information is stored. A stateless proxy never retransmits a message, and does not use any SIP timers. Note that the stateless loop detection using Via header fields described in RFC 2543 has been deprecated (removed) in RFC 3261 in favor of the use of a mandatory Max-Forwards header field in all requests.

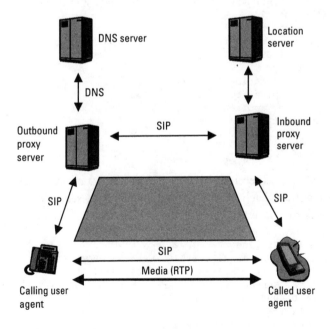

Figure 3.3 SIP Trapezoid.

A stateful proxy server keeps track of requests and responses received in the past and uses that information in processing future requests and responses. For example, a stateful proxy server starts a timer when a request is forwarded. If no response to the request is received within the timer period, the proxy will retransmit the request, relieving the user agent of this task. Also, a stateful proxy can require user agent authentication, as described in Section 3.8.

The most common type of SIP proxy is a transaction stateful proxy. A transaction stateful proxy keeps state about a transaction but only for the duration that the request is pending. For example, a transaction stateful proxy would keep state when it receives an INVITE request until it received a 200 OK or a final failure response (e.g., 404 Not Found). After that, it would destroy the state information. This allows a proxy to perform useful search services but minimize the amount of state storage required.

One such example of a search service is a proxy server that receives an INVITE request, then forwards it to a number of locations at the same time. This "forking" proxy server keeps track of each of the outstanding requests and the response to each, as shown in Figure 3.4. This is useful if the location service or database lookup returns multiple possible locations for the called party that need to be tried.

In the example of Figure 3.4, the INVITE contains:

```
INVITE sip:support@chaos.info SIP/2.0
Via: SIP/2.0/UDP 45.2.32.1:5060 ;branch=z9hG4bK67865
Max-Forwards: 70
To: <sip:support@chaos.info>
From: A. N. Sarkovskii <sip:sarkovskii@45.2.32.1>;tag=7643545
Call-ID: 0140092501
CSeq: 1 INVITE
Subject: Bifurcation Question
Contact: <sip:sarkovskii@45.2.32.1>
Content-Type: application/sdp
Content-Length: ...

(SDP not shown)
```

The INVITE is received by the chaos.info proxy server, which forks to two user agents. Each user agent begins alerting, sending two provisional responses back to Sarkovskii. They are:

```
SIP/2.0 180 Ringing
Via: SIP/2.0/UDP 45.2.32.1:5060;branch=z9hG4bK67865
To: <sip:support@chaos.info>;tag=343214112
From: A. N. Sarkovskii <sip:sarkovskii@45.2.32.1>;tag=7643545
Call-ID: 0140092501
CSeq: 1 INVITE
Contact: <sip:agent42@67.42.2.1>
Content-Length: 0
```

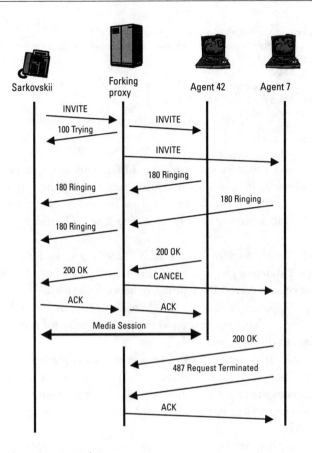

Figure 3.4 Forking proxy operation.

and:

```
SIP/2.0 180 Ringing
Via: SIP/2.0/UDP 45.2.32.1:5060;branch=z9hG4bK67865
To: <sip:support@chaos.info>;tag=a5ff34d9ee201
From: A. N. Sarkovskii <sip:sarkovskii@45.2.32.1>;tag=7643545
Call-ID: 0140092501
CSeq: 1 INVITE
Contact: <sip:agent7@67.42.2.32>
Content-Length: 0
```

The two responses are identical except for having different To tags and Contact URIs. Finally, one of the two UAs answers and sends a 200 OK response:

```
SIP/2.0 200 OK
Via: SIP/2.0/UDP 45.2.32.1:5060;branch=z9hG4bK67865
```

```
To: <sip:support@chaos.info>;tag=343214112
From: A. N. Sarkovskii <sip:sarkovskii@45.2.32.1>;tag=7643545
Call-ID: 0140092501
CSeq: 1 INVITE
Contact: <sip:agent42@67.42.2.1>
Content-Type: application/sdp
Content-Length: ...
```

(SDP not shown)

The forking proxy server sends a CANCEL to the second UA to stop that phone alerting. If both UAs had answered, the forking proxy would have forwarded both 200 OK responses back to the caller who then would have had to choose which one, probably accepting one and sending a BYE to the other one.

A stateful proxy usually sends a 100 Trying response when it receives an INVITE. A stateless proxy never sends a 100 Trying response. A 100 Trying response received by a proxy is never forwarded—it is a single hop only response. A proxy handling a TCP request must be stateful, since a user agent will assume reliable transport and rely on the proxy for retransmissions on any UDP hops in the signaling path.[1]

The only limit to the number of proxies that can forward a message is controlled by the Max-Forwards header field, which is decremented by each proxy that touches the request. If the Max-Forwards count goes to zero, the proxy discards the message and sends a 483 Too Many Hops response back to the originator.

A SIP session timer [8] has been proposed to limit the time period over which a stateful proxy must maintain state information. In the initial INVITE request, a Session-Expires header field indicates a timer interval after which stateful proxies may discard state information about the session. User agents must tear down the call after the expiration of the timer. The caller can send re-INVITEs to refresh the timer, enabling a "keep alive" mechanism for SIP. This solves the problem of how long to store state information in cases where a BYE request is lost or misdirected, or in other security cases described in later sections. The details of this implementation are described in Section 6.2.29.

3.5.2 Redirect Servers

A redirect server was introduced in Figure 2.6 as a type of SIP server that responds to, but does not forward requests. Like a proxy sever, a redirect server

---

1. TCP usually provides end-to-end reliability for applications. In SIP, however, TCP only provides single-hop reliability. End-to-end reliability is only achieved by a chain of TCP hops or TCP hops interleaved with UDP hops and stateful proxies.

uses a database or location service to look up a user. The location information, however, is sent back to the caller in a redirection class response (3xx), which, after the ACK, concludes the transaction. Figure 3.5 shows a call flow that is very similar to the example of Figure 2.2, except the server uses redirection instead of proxying to assist Schroedinger locate Heisenberg.

The INVITE from Figure 3.5 contains:

```
INVITE sip:werner.heisenberg@munich.de SIP/2.0
Via: SIP/2.0/UDP 100.101.102.103:5060 ;branch=z9hG4bK54532
Max-Forwards: 70
To: Heisenberg <sip:werner.heisenberg@munich.de>
From: E. Schroedinger <sip:schroed5244@aol.com>;tag=4313413
Call-ID: 9@100.101.102.103
CSeq: 1 INVITE
Subject: Where are you exactly?
Contact: <sip:schroed5244@pc33.aol.com>
Content-Type: application/sdp
Content-Length: 159

v=0
o=schroed5244 2890844526 2890844526 IN IP4 100.101.102.103
s=Phone Call
t=0 0
c=IN IP4 100.101.102.103
m=audio 49172 RTP/AVP 0
a=rtpmap:0 PCMU/8000
```

The redirection response to the INVITE is sent by the redirect server:

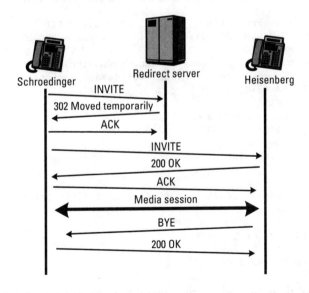

Figure 3.5 Example with redirect server.

```
SIP/2.0 302 Moved Temporarily
Via: SIP/2.0/UDP 100.101.102.103:5060;branch=z9hG4bK54532
To: Heisenberg <sip:werner.heisenberg@munich.de>;tag=052500
From: E. Schroedinger <sip:schroed5244@aol.com>;tag=4313413
Call-ID: 9@100.101.102.103
CSeq: 1 INVITE
Contact: sip:werner.heisenberg@200.201.202.203
Content-Length: 0
```

Schroedinger acknowledges the response:

```
ACK sip:werner.heisenberg@munich.de SIP/2.0
Via: SIP/2.0/UDP 100.101.102.103:5060;branch=z9hG4bK54532
Max-Forwards: 70
To: Heisenberg <sip:werner.heisenberg@munich.de>;tag=052500
From: E. Schroedinger <sip:schroed5244@aol.com>;tag=4313413
Call-ID: 9@100.101.102.103
CSeq: 1 ACK
Content-Length: 0
```

Notice that the ACK request reuses the same branch ID as the INVITE and the 302 response. This is because an ACK to a non-2xx final response is considered to be part of the same transaction as the INVITE. Only an ACK sent in response to a 200 OK is considered a separate transaction with a unique branch ID. Also, an ACK to a non-2xx final response is a hop-by-hop response, not an end-to-end response as discussed in Section 3.6.

This exchange completes this call attempt, so a new INVITE is generated with a new Call-ID and sent directly to the location obtained from the Contact header field in the 302 response from the redirect server:

```
INVITE sip:werner.heisenberg@200.201.202.203 SIP/2.0
Via: SIP/2.0/UDP 100.101.102.103:5060;branch=z9hG4bK92313
Max-Forwards: 70
To: Heisenberg <sip:werner.heisenberg@munich.de>
From: E. Schroedinger <sip:schroed5244@aol.com>;tag=13473
Call-ID: 54-67-45-23-13
CSeq: 1 INVITE
Subject: Where are you exactly?
Contact: <sip:schroed5244@pc33.aol.com>
Content-Type: application/sdp
Content-Length: 159

v=0
o=schroed5244 2890844526 2890844526 IN IP4 100.101.102.103
s=Phone Call
t=0 0
c=IN IP4 100.101.102.103
m=audio 49172 RTP/AVP 0
a=rtpmap:0 PCMU/8000
```

The call then proceeds in the same way as Figure 2.2, with the messages being identical. Note that in Figure 3.5, a 180 Ringing response is not sent;

instead, the 200 OK response is sent right away. Since 1xx informational responses are optional, this is a perfectly valid response by the UAS if Heisenberg responded to the alerting immediately and accepted the call. In the PSTN, this scenario is called fast answer.

3.5.3 Registration Servers

A SIP registration server was introduced in the example of Figure 2.3. A registration server, also known as a registrar, accepts SIP REGISTER requests; all other requests receive a 501 Not Implemented response. The contact information from the request is then made available to other SIP servers within the same administrative domain, such as proxies and redirect servers. In a registration request, the To header field contains the name of the resource being registered, and the Contact header fields contain the alternative addresses or aliases. The registration server creates a temporary binding between the Address of Record (AOR) URI in the To and the device URI in the Contact Courier.

Registration servers usually require the registering user agent to be authenticated, using means described in Section 3.8, so that incoming calls cannot be hijacked by an unauthorized user. This could be accomplished by an unauthorized user registering someone else's SIP URI to point to their own phone. Incoming calls to that URI would then ring the wrong phone. Depending on the header fields present, a REGISTER request can be used by a user agent to retrieve a list of current registrations, clear all registrations, or add a registration URI to the list. These types of requests are described in Section 4.1.2.

For full registration security, TLS must be used as HTTP Digest does not provide the needed integrity protection.

3.6 Acknowledgment of Messages

Most SIP requests are end-to-end messages between user agents. That is, proxies between the two user agents simply forward the messages they receive and rely on the user agents to generate acknowledgments or responses.

There are some exceptions to this general rule. The CANCEL method (used to terminate pending calls or searches and discussed in detail in Section 4.1.5) is a hop-by-hop request. A proxy receiving a CANCEL immediately sends a 200 OK response back to the sender and generates a new CANCEL, which is then forwarded to the next hop. (The order of sending the 200 OK and forwarding the CANCEL is not important.) This is shown in Figure 4.4.

Other exceptions to this rule include 3xx, 4xx, 5xx, and 6xx responses to an INVITE request. While an ACK to a 2xx response is generated by the end point, a 3xx, 4xx, 5xx, or 6xx response is acknowledged on a hop-by-hop basis. A proxy server receiving one of these responses immediately generates an ACK back to the sender and forwards the response to the next hop. This type of hop-by-hop acknowledgment is shown in Figure 4.2.

ACK messages are only sent to acknowledge responses to INVITE requests. For responses to all other request types, there is no acknowledgment. A lost response is detected by the UAS when the request is retransmitted.

3.7 Reliability

SIP has reliability mechanisms defined, which allow the use of unreliable transport layer protocols such as UDP. When SIP uses TCP or TLS, these mechanisms are not used, since it is assumed that TCP will retransmit the message if it is lost and inform the client if the server is unreachable.

For SIP transport using UDP, there is always the possibility of messages being lost or even received out of sequence, because UDP guarantees only that the datagram is error-free. A UAS validates and parses a SIP request to make sure that the UAC has not errored by creating a request missing required header fields or other syntax violations. Reliability mechanisms in SIP include:

- Retransmission timers;
- Increasing command sequence CSeq numbers;
- Positive acknowledgments.

SIP timer T1 is started by a UAC or a stateful proxy server when a new request is generated or sent. If no response to the request (as identified by a response containing the identical local tag, remote tag, Call-ID, and CSeq) is received when T1 expires, the request is re-sent. If a provisional (informational class 1xx) response is received, the UAC or stateful proxy server ignores T1 and starts a new longer timer T2. No retransmissions are sent until T2 expires.

After a request is retransmitted, the timer period is doubled until T2 is reached. After that, the remaining retransmissions occur at T2 intervals. This capped exponential backoff process is continued until a maximum of 10 retransmissions at increasing intervals are sent. A stateful proxy server that receives a retransmission of a request discards the retransmission and continues its retransmission schedule based on its own timers. Typically, it will resend the last provisional response.

For an INVITE request, the retransmission scheme is slightly different. After a provisional (1xx) response is received, the INVITE is never retransmitted. However, a proxy may discard transaction state after 3 minutes.

A stateful proxy must store a forwarded request or generated response message for 32 seconds. An example message flow involving two user agents, a stateful proxy, two lost messages (shown by "X" in Figure 3.6), and three retransmissions is shown in Figure 3.6. In this example, the OPTIONS sent by the stateful proxy server to the UAS is lost. As a result, it is retransmitted when the proxy's T1 timer expires and no response is received. The 200 OK forwarded by the proxy is also lost. When timer T1 in the UAC expires without a final response (non 1xx response), the OPTIONS is retransmitted. When the proxy receives the retransmitted OPTIONS, it deduces that the 200 OK was lost and resends it. The proxy recognizes the 200 OK as a retransmission and does not forward it.

Suggested default values for T1 and T2 are 500 ms and 4 seconds, respectively. Timer T1 is supposed to be an estimate of the round-trip time (RTT) in the network. Longer values are allowed but not shorter ones, because this will generate more message retransmissions. See Table 4 in RFC 3261 [1] for a summary of SIP timers.

Note that gaps in CSeq number do not always indicate a lost message. In the authentication examples in the next section, not every request (and hence CSeq) generated by the UAC will reach the UAS if authentication challenges occur by proxies in the path.

3.8 Authentication

Authentication in SIP takes two general forms. One is the authentication of a user agent by a proxy, redirect, or registration server. The other is the

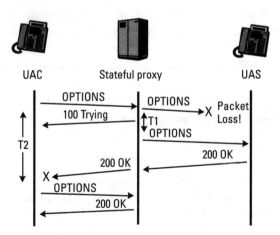

Figure 3.6 SIP reliability example.

authentication of a user agent by another user agent. Mutual authentication between proxies or a proxy and a user agent is also possible using certificates.

A proxy or redirect server might require authentication to allow a user agent to access a service or feature. For example, a proxy server may require authentication before forwarding an INVITE to a gateway or invoking a service. A registration server may require authentication to prevent incoming call hijacking as described previously. User agents can authenticate each other to verify who they are communicating with, since From header fields are easily forged.[2]

SIP supports both a simple lightweight authentication scheme and a very robust scheme. The simple scheme is based on HTTP Digest [9] and uses a simple challenge/response mechanism and a shared secret between the two servers. The other schemes involve cryptographic means to exchange and verify certificates.

Using HTTP Digest authentication, a proxy requiring authentication replies to an unauthenticated INVITE with a 407 Proxy Authorization Required response containing a Proxy-Authenticate header field with the form of the challenge. After sending an ACK for the 407, the user agent can then resend the INVITE with a Proxy-Authorization header field containing the credentials. This request is usually sent using the same Call-ID but an incremented CSeq count. User agent, redirect, or registrar servers typically use 401 Unauthorized response to challenge authentication containing a WWW-Authenticate header field, and expect the a new INVITE to contain an Authorization header field containing the user agent's credentials. A user agent's credentials are usually an encrypted username and password, as in the example of Section 10.1. The Authentication-Info header field also allows a user agent to authenticate a proxy server using HTTP Digest. However, HTTP Digest does not provide integrity protection—for this, TLS or S/MIME must be used.

A call flow involving both proxy and user agent authentication is shown in Figure 3.7.

If TLS is used as transport, as discussed in Section 2.5.3, the two servers can exchange and verify certificates for authentication.

---

2. The ability to forge From header fields is present in SMTP, where it is virtually a *feature*. A preference setting in an e-mail program sets your name and e-mail address, which need not correspond to the address or domain that is used to send the message. This allows a user to send multiple e-mail addresses from the same e-mail account by simply changing the From address before sending a message. Only a detailed examination of a full set of SMTP header fields will show that the e-mail was sent from another address.

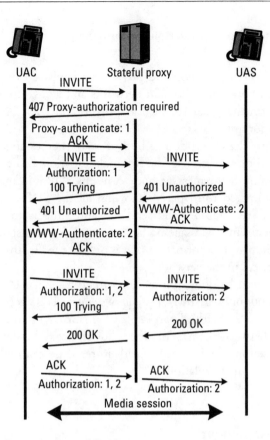

Figure 3.7 Digest authentication call flow.

3.9 S/MIME Encryption

While authentication is used as a means of access control and identity confirmation, encryption is used for privacy. SIP messages intercepted during session setup reveal considerable information, including:

- Both parties' SIP URIs and IP addresses;
- The fact that the two parties have established a call;
- The IP addresses and port numbers associated with the media, allowing eavesdropping.

Presence information can reveal even more private information, such as:

- The user's geographic location;
- The user's current activity level;

- The "mood" or other personal information;

As such, strong encryption and privacy mechanisms have been built into SIP that work in both single-domain networks and across the public Internet. The use of intermediary devices such as proxies and B2BUAs also make SIP security very important.

SIP supports the encryption of both message bodies and message header fields. The encryption of message bodies makes it more difficult for an eavesdropper to listen in, if the message body contains SDP information, for example. Also, an uninvited third party, knowing all the SDP information could guess the RTP SSRC number and send unwanted media to either party, so-called media spamming. Message bodies in presence and instant message messages also contain private information that is only needed by the user agents, not intermediary devices in the network.

Encryption can be done hop-by-hop or end-to-end. TLS can be used for the hop-by-hop but only S/MIME [10] allows end-to-end encryption. S/MIME allows UAs to discover if a third part (a proxy or B2BUA) is modifying SIP messages or bodies between them. Note that allowed proxy header field modification (such as deletion of Via header fields in response routing) is allowed and not considered a security violation.

3.10 Multicast Support

SIP support for UDP multicast has been mentioned in previous sections. There are two main uses for multicast in SIP.

SIP registration can be done using multicast, by sending the REGISTER message to the well-known "All SIP Servers" URI sip:sip.mcast.net at IP address 224.0.1.75.

The second use for multicast is to send a multicast session invitation. This effectively allows a conference call to be established with a single request. An INVITE with a partially defined Request-URI can be sent using multicast. For example, a multicast INVITE could be sent to sip:*@mci.com, which would invite all MCI employees with SIP phones receiving the request to respond. Responses to a multicast request are also sent by multicast. To limit congestion, only a limited set of responses is allowed by the standard.

A proxy can forward a unicast INVITE request to a multicast address. The use of multicast is recorded in a SIP message using the maddr parameter in the Via header field, as discussed in Section 6.1.14.

However, due to the limited implementations of multicast, these features are rarely used.

3.11   Firewalls and NAT Interaction

Most corporate LANs or intranets connect to the public Internet through a firewall. A firewall is filtering software usually in a router or hub that is used to protect the LAN behind it from various kinds of attacks and unauthorized access. Firewalls are also increasingly being used in home network routers and wireless hubs and in PCs themselves. Sometimes they are used to prevent users behind the firewall accessing certain resources in the Internet. In the simplest deployment, a firewall can be thought of as a one-way gate: It allows outgoing packets from the intranet to the Internet, but blocks incoming packets from the Internet unless they are responses to queries. Only certain types of requests from the Internet will be allowed to pass through the firewall, such as HTTP requests to the corporate Web server, SMTP e-mail messages, or DNS queries to the authoritative DNS for the corporate domain. The firewall does this by keeping track of TCP connections opened and filtering ports.

Firewalls pose a particularly difficult challenge to SIP sessions. Because SIP can use TCP and a well-known port, configuring a firewall to pass SIP is not too difficult. This does not help the media path, however, which uses RTP over UDP on various ports and will be blocked by most firewalls. A firewall or a proxy that controls the firewall needs to understand SIP, be able to parse an INVITE request and 200 OK response, extract the IP addresses and port numbers from the SDP, and open up "pin holes" in the firewall to allow this traffic to pass. The hole can then be closed when a BYE is sent or a session timer expires. An alternative is an ALG—a B2BUA that is trusted by the firewall. The firewall then allows SIP and RTP traffic, which terminates on the ALG and blocks all other traffic. The authentication and security policies of allowing or denying SIP sessions are then controlled by the SIP ALG instead of in the firewall itself.

Network address translators (NATs) also cause serious problems for SIP. A NAT can be used to conserve IPv4 addresses, or can be used to hide the IP address and LAN structure behind the NAT. It is used on a router or firewall that provides the only connection of a LAN to the Internet, a so-called stub network. A NAT allows nonunique IP addresses to be used internally within the LAN. When a packet is sent from the LAN to the Internet, the NAT changes the nonglobally unique address (usually addresses in the range 10.x.x.x, 172.16.x.x – 172.29.x.x and 192.168.x.x) in the packet header to a globally unique address from a pool of available addresses. Addresses can also be statically assigned. This means that every node on the network does not have to have a globally unique IP address. Responses from the Internet are translated back to the nonunique address. A NAT, however, is not completely transparent to higher layers. For a signaling protocol such as SIP, a NAT can cause particular problems.

Because responses in SIP are routed using `Via` header fields, a device behind a NAT will stamp its nonroutable private IP address in its `Via` header field of messages that it originates. When the request is forwarded outside the intranet by the NAT, the UDP and IP packet headers will be rewritten with a temporarily assigned global Internet address. The NAT will keep track of the *binding* between the local address and the global address so that incoming packets can have the UDP and IP headers rewritten and routed correctly. However, IP addresses in a SIP message, such as `Via` and `Contact` header fields, or IP addresses in SDP message bodies will not be rewritten and will not be routable.

To partially solve the message routing problem, SIP has a mechanism for detecting if a NAT is present in a SIP message path. Each proxy or user agent that receives a request checks the received IP address with the address in the `Via` header field. If the addresses are different, there is a NAT between them. The unroutable `Via` header field is fixed with a `received` tag containing the actual global IP address. Outside the NAT, the response is routed using the received IP address. Inside the NAT, the `Via` address is used. This does solve the message response routing problem (except when the port number is also wrong), but not the media problems.

Another problem with NATs is the time span of the NAT address binding. For a TCP connection, this is not an issue—the binding is maintained as long as the connection is open. For a UDP SIP session, the time period is determined by the application. If a binding were removed before a BYE was sent terminating the session, the connection would effectively be closed and future signaling impossible. A keep-alive mechanism may be needed to refresh this binding.

A SIP ALG coresident with the NAT solves many of these problems. The ALG would rewrite the media IP addresses in the SIP messages and would not allow the NAT to remove the address binding until a BYE was sent or a session timer had expired. However, NATs are often deployed deep inside a service provider's network that is not associated with providing SIP service, and hence has no incentive to upgrading the NATs to allow this service to work.

Even without ALGs or upgrades in NATs, it is possible to use SIP to establish a media session through many types of NATs. The protocols described in the next section allow a SIP client to discover the presence and type of NATs between it and the public Internet, learn its public IP address, and possibly fix the incorrect addresses in the SIP and SDP messages.

3.12   Protocols and Extensions for NAT Traversal

A detailed analysis of various scenarios and solutions to NAT traversal has been done [11]. Some of the results are summarized in this section along with two non-SIP protocols: Simple Traversal of UDP through NATs (STUN) and Traversal Using Relay NAT (TURN), which aid in NAT traversal. Finally, some extensions to SIP and SDP to enable NAT traversal are discussed.

3.12.1 STUN Protocol

The STUN protocol [12] allows a client to discover the presence and type of NATs between the client and the public Internet. In addition, a client can discover the mapping between the private IP address and port number and the public IP address and port number. Typically, a service provider will operate a STUN server in the public Internet, with STUN clients being embedded in end-devices, which are possibly behind a NAT.

A STUN server can be located using DNS SRV records using the service provider's domain as the lookup. STUN typically uses the well-known port number 3478. STUN is a binary encoded protocol with a 20-octet header field and possibly additional attributes.

Since a STUN client uses the protocol to learn public IP addresses, some security is necessary or an imposter STUN server could provide incorrect public IP addresses and block or intercept communication destined for the client. As a result, the first step with STUN is for the client to contact the server and negotiate a shared secret (usually a username and password) over a secure link, which will be used in future STUN requests. The initial STUN connection is opened using TLS over TCP. The client verifies the certificate of the server to make sure that it is connected to the proper server. The client then sends a Shared Secret Request packet to the server and receives the shared secret to be used.

The client is then ready to use STUN to determine the presence and type of NATs between the client and the server. Using the shared secret, the client sends one or more Binding Request packets using UDP to the server. These packets must be sent from the same IP address that the client will use for the other protocol, since this is the address translation information that the client is trying to discover. The server returns Binding Response packets, which tell the client the public IP address and port number from which it received the Binding Request. Since the client knows the private IP address and port number it used to send the Binding Request, it learns the mapping between the private and public address space being performed by the NAT. Of course, if the Binding Response packets indicate the same address and port number as the request, this tells the client that no NATs are present.

Figure 3.8 shows a SIP client behind a NAT obtaining its public IP address using STUN and setting up a session as a result.

The type of NAT can be determined by sending multiple Binding Request packets with different attributes. As a result the scenario type can be determined

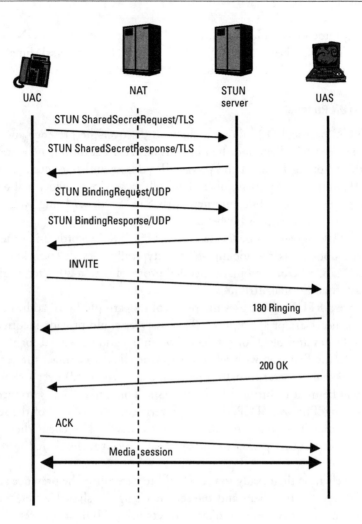

Figure 3.8 SIP and STUN for NAT traversal call flow.

to be one of those listed in Table 3.1. STUN can also be used to refresh the NAT address binding to keep it valid during the application layer session.

For all the cases described in Table 3.1 except the symmetric NAT, STUN provides the client with the public IP addresses needed to establish a media session using SIP. The IP addresses obtained using STUN are used in the Via, Contact URI and in the SDP media information in the INVITE, for example, and result in a successful media session with a client in the public Internet.

However, to traverse a symmetric NAT, or for certain topologies such as communication between two user agents behind NATs, a signaling and media relay protocol such as TURN is required, which is described in the next section.

Table 3.1
Type of Address Translation

Type	Translation
Internet	No translation; client has a public IP address
Full cone NAT	Constant mapping between private IP address and public IP address
Restricted cone NAT	Constant mapping, but an outgoing packet is needed to open incoming path
Symmetric NAT	Different public IP address mapping is used based on destination IP address

3.12.2 TURN Protocol

The TURN protocol [13] allows a client to obtain a transport IP address and port that it can receive packets sent from a single IP address in the public Internet. For some NAT topologies such as a client behind a symmetric NAT, using a relay located in the public Internet is the only approach that allows communication to take place.

Similar to STUN, a TURN client can use DNS SRV records for the domain of the service provider. TURN uses an identical syntax to STUN and reuses the Shared Secret Request and Shared Secret Response packets to establish the shared secret. The client can also use Binding Request and Binding Response packets to detect and categorize the NAT in the path. A client uses an Allocate Request to request a relay IP address and port number be returned in an Allocate Response packet.

Requests for transport addresses to be used for UDP must be sent using UDP while requests for transport addresses for TCP transport must be sent using TCP.

TURN, due to its triangular routing, will result in increased packet latency and increased probability of packet loss. TURN should only be used when it is the only approach available (i.e., the IP addresses obtained using STUN will not work). Note that TURN can be used instead of a B2BUA to provide an anonymizer service.

Figure 3.9 shows a SIP client behind a NAT using STUN and TURN to obtain a transport IP address and setting up a session as a result. Note that the resulting SIP and RTP messages are routed through the TURN server.

In some cases, the use of STUN and TURN may result in the use of a relay when, in fact, the two clients can communicate successfully without one. This case is where both clients are behind the same NAT.

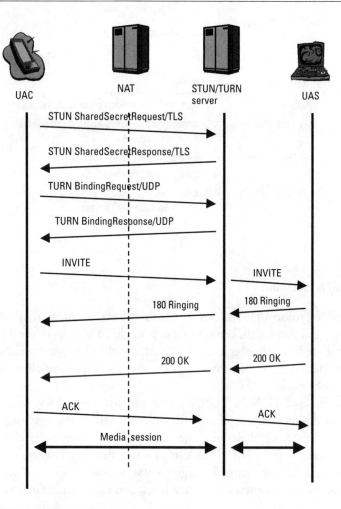

Figure 3.9 SIP and TURN for NAT traversal call flow.

3.12.3 Other SIP/SDP NAT-Related Extensions

Other SIP and SDP extensions have been developed to solve some of the problems associated with NATs. They are described in the following sections and can be used in conjunction with protocols such as STUN and TURN to enable NAT traversal. Each extension solves a piece but not the complete NAT traversal problem. Instead, they represent an optimization or improvement used along with STUN and TURN.

As mentioned previously, using TCP transport for SIP makes NAT traversal easier. In particular, it is easier to keep the private/public address binding for a TCP connection than it is with UDP. This is a particular problem for registrations. If UDP is used for registration behind a NAT, either STUN packets or repeated registrations are needed (as often as every minute). It is better to register using TCP, but only if the TCP connection can be kept open and incoming requests are routed over that open connection. A SIP extension has been proposed to solve this problem [14].

An example `Via` header field in a `REGISTER` request contains a parameter `rport`, which indicates that the UAC supports this extension:

```
Via: SIP/2.0/UDP client.behind.nat.org;rport;branch=z9hG4bK92313s
```

The next hop server that supports the extension would record both the received IP address and port number in the `Via` header field; for example:

```
Via: SIP/2.0/UDP client.behind.nat.org;rport=23131
   ;received=192.0.1.2;branch=z9hG4bK92313s
```

This TCP connection would then be kept open and responses to this user would be sent back to the client at address `192.0.1.2` port `23131`. An incoming `INVITE` or other request would be routed over this open TCP connection.

The ability to specify in SDP a "symmetric" RTP session [15] has been proposed as a solution for some NAT traversal problems. Since a client behind a NAT can usually successfully send RTP packets to another client in the public Internet, in a symmetric mode, RTP sent in the other direction could be sent to the address and port that RTP was received from. For example, consider an SDP offer in an `INVITE` sent by the UAC which supports this connection-oriented media extension:

```
v=0
o=client 28908445312 28908445312 IN IP4 10.1.2.23
s=-
t=0 0
c=IN IP4 10.1.2.23
m=audio 49172 RTP/AVP 0
a=rtpmap:0 PCMU/8000
a=direction:active IN IP4
```

If the UAS supports this extension, it will wait for RTP packets to be received from the client behind the NAT before sending. The answer SDP will be:

```
v=0
o=client 28908445214 28908445214 IN IP4 client.public.org
s=-
t=0 0
c=IN IP4 client.public.org
m=audio 54332 RTP/AVP 0
a=rtpmap:0 PCMU/8000
```

```
a=direction:passive IN IP4
```

The direction attribute in the answer tells the UAC that the UAS supports the symmetric RTP extension. The UAS will then send its RTP packets to the IP address and port number received from the UAC which overrides the unreachable 10.1.2.23 address and 49172 port number in the SDP.

Finally, it is assumed that RTP Control Protocol (RTCP) packets (see Section 7.2) are received on one port higher than the RTP port number. If STUN or TURN is used to obtain port numbers, this consecutive numbering may not be possible if the NAT is mapping port numbers as well. Also, RTCP may need to be sent to an entirely different IP address. A solution to this is an explicit RTCP port number and possibly IP address in SDP extension [16]. This allows RTCP to still be received in these cases. For example, a UAC behind a NAT could include the attribute:

```
a=rtcp:53020 IN IP4 126.16.64.4
```

A UAS supporting the extension would send RTCP packets to this IP address and port number instead of using the RTP IP address (connection line) and port number plus one. If just the port number needs to be changed, the address information can be omitted:

```
a=rtcp:53022
```

Note that without this SDP extension, the SIP and RTP sessions will still work, but no RTCP would be exchanged.

References

[1]   Rosenberg, J., et al., "SIP: Session Initiation Protocol," RFC 3261, 2002.

[2]   Rosenberg, J., "A Presence Event Package for the Session Initiation Protocol (SIP)," IETF Internet-Draft, Work in Progress, January 2003.

[3]   Roach, A., "Session Initiation Protocol (SIP)-Specific Event Notification," RFC 3265, 2002.

[4]   Campbell, B., et al., "SIMPLE Presence Publication Mechanism," IETF Internet-Draft, Work in Progress, February 2003.

[5]   Schulzrinne, H., and C. Agboh, "Session Initiation Protocol (SIP)-H.323 Interworking Requirements," IETF Internet-Draft, Work in Progress, February 2003.

[6]   Rosenberg, J., H. Salama, and M. Squire, "Telephony Routing over IP (TRIP)," RFC 3219, 2002.

[7]   Bangalore, M., et al., "A Telephony Gateway REgistration Protocol (TGREP)," IETF Internet-Draft, February 2003.

[8]   Donovan, S., and J. Rosenberg, "The SIP Session Timer," IETF Internet-Draft, Work in Progress.

[9]   Franks, J., et al., "HTTP Authentication: Basic and Digest Access Authentication," RFC 2617, 1999.

[10] Ramsdell, B., "S/MIME Version 3 Message Specification," RFC 2633, 1999.

[11] Rosenberg, J., R. Mahy, and S. Sen, "NAT and Firewall Scenarios and Solutions for SIP," IETF Internet-Draft, Work in Progress, March 2003.

[12] Rosenberg, J., et al., "STUN—Simple Traversal of User Datagram Protocol (UDP) Through Network Address Translators (NATs)," RFC 3489, 2003.

[13] Rosenberg, J., et al., "Traversal Using Relay NAT (TURN)," IETF Internet-Draft, Work in Progress, March 2003.

[14] Yon, D., "Connection-Oriented Media Transport in SDP," IETF Internet-Draft, Work in Progress, May 2002.

[15] Rosenberg, J., J. Weinberger, and H. Schulzrinne, "An Extension to the Session Initiation Protocol (SIP) for Symmetric Response Routing," RFC 3581, 2003.

[16] Huitema, C., "RTCP Attribute in SDP," IETF Internet-Draft, Work in Progress, September 2002.

4

SIP Request Messages

This chapter covers the types of SIP requests called methods. Six are described in the SIP specification document RFC 3261 [1]. Seven more methods are described in separate RFC documents. After discussing the methods, this chapter concludes with a discussion of SIP URLs and URIs, tags, and message bodies.

4.1 Methods

SIP requests or methods are considered "verbs" in the protocol, since they request a specific action to be taken by another user agent or server. The INVITE, REGISTER, BYE, ACK, CANCEL, and OPTIONS methods are the original six methods in SIP. The REFER, SUBSCRIBE, NOTIFY, MESSAGE, UPDATE, INFO, and PRACK methods are described in separate RFCs.

Note that a proxy does not need to understand a request method in order to forward the request. A proxy treats an unknown method as if it were an OPTIONS; that is, it forwards the request to the destination if it can. This allows new features and methods useful for user agents to be introduced without requiring support from proxies that may be in the middle. A user agent receiving a method it does not support replies with a 501 Not Implemented response. Method names are case sensitive and conventionally use all uppercase for visual clarity to distinguish them from header fields, which use both upper and lower case.

4.1.1 INVITE

The INVITE method is used to establish media sessions between user agents. In telephony, it is similar to a Setup message in ISDN or an initial address message (IAM) in ISUP. (PSTN protocols are briefly introduced in Section 7.4.) Responses to INVITEs are always acknowledged with the ACK method described in Section 4.1.4. Examples of the use of the INVITE method are described in Chapter 2.

An INVITE usually has a message body containing the media information of the caller. The message body can also contain other session information such as quality of service (QoS) or security information. If an INVITE does not contain media information, the ACK contains the media information of the UAC. An example of this call flow is shown in Figure 4.1. If the media information contained in the ACK is not acceptable, then the called party must send a BYE to cancel the session—a CANCEL cannot be sent because the session is already established. A media session is considered established when the INVITE, 200 OK, and ACK messages have been exchanged between the UAC and the UAS. A successful INVITE request establishes a dialog between the two user agents, which continues until a BYE is sent by either party to end the session, as described in Section 4.1.3.

A UAC that originates an INVITE to establish a dialog creates a globally unique Call-ID that is used for the duration of the call. A CSeq count is initialized (which need not be set to 1, but must be an integer) and incremented for each new request for the same Call-ID. The To and From headers are populated with the remote and local addresses. A From tag is included in the

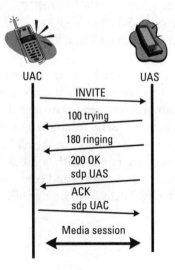

Figure 4.1 INVITE with no SDP message body.

INVITE, and the UAS includes a To tag in any responses, as described in Section 4.3. A To tag in a 200 OK response to an INVITE is used in the To header field of the ACK and all future requests within the dialog. The combination of the To tag, From tag, and Call-ID is the unique identifier for the dialog.

An INVITE sent for an existing dialog references the same Call-ID as the original INVITE and contains the same To and From tags. Sometimes called a re-INVITE, the request is used to change the session characteristics or refresh the state of the dialog. The CSeq command sequence number is incremented so that a UAS can distinguish the re-INVITE from a retransmission of the original INVITE.

If a re-INVITE is refused or fails in any way, the session continues as if the INVITE had never been sent. A re-INVITE must not be sent by a UAC until a final response to the initial INVITE has been received—instead, an UPDATE request can be sent, as described in Section 4.1.13. There is an additional case where two user agents simultaneously send re-INVITEs to each other. This is handled in the same way with a Retry-After header. This condition is called *glare* in telephony, and occurs when both ends of a trunk group seize the same trunk at the same time.

An Expires header in an INVITE indicates to the UAS how long the call request is valid. For example, the UAS could leave an unanswered INVITE request displayed on a screen for the duration of specified in the Expires header. Once a session is established, the Expires header has no meaning—the expiration of the time does not terminate the media session. Instead, a Session-Expires header (Section 6.2.26) can be used to place a time limit on an established session.

An example INVITE request with a SDP message body is shown below:

```
INVITE sip:411@salzburg.at;user=phone SIP/2.0
Via: SIP/2.0/UDP salzburg.edu.at:5060;branch=z9hG4bK1d32hr4
Max-Forwards:70
To: <sip:411@salzburg.at;user=phone>
From: Christian Doppler <sip:c.doppler@salzburg.edu.at>
  ;tag=817234
Call-ID: 12-45-A5-46-F5@salzburg.edu.at
CSeq: 1 INVITE
Subject: Train Timetables
Contact: sip:c.doppler@salzburg.edu.at
Content-Type: application/sdp
Content-Length: 151

v=0
o=doppler 2890842326 2890844532 IN IP4 salzburg.edu.at
s=Phone Call
c=IN IP4 50.61.72.83
t=0 0
```

```
m=audio 49172 RTP/AVP 0
a=rtpmap:0 PCMU/8000
```

In addition to the required headers, this request contains the optional `Subject` header. Note that this Request-URI contains a phone number. Phone number support in SIP URIs is described in Section 4.2.

The mandatory and header field in an `INVITE` request are shown in Table 4.1.

4.1.2 REGISTER

The `REGISTER` method is used by a user agent to notify a SIP network of its current `Contact` URI (IP address) and the URI that should have requests routed to this `Contact`. As mentioned in Section 2.3, SIP registration bears some similarity to cell phone registration on initialization. Registration is not required to enable a user agent to use a proxy server for outgoing calls. It is necessary, however, for a user agent to register to receive incoming calls from proxies that serve that domain unless some non-SIP mechanism is used by the location service to populate the SIP URIs and `Contacts` of end-points. A `REGISTER` request may contain a message body, although its use is not defined in the standard. Depending on the use of the `Contact` and `Expires` headers in the `REGISTER` request, the registrar server will take different action. Examples of this are shown in Table 4.2. If no `expires` parameter or `Expires` header is present, a SIP URI will expire in 1 hour. The presence of an `Expires` header sets the expiration of `Contacts` with no `expires` parameter. If an `expires` parameter is present, it sets the expiration time for that `Contact` only. Non-SIP URIs have no default expiration time.

The `CSeq` is incremented for a `REGISTER` request. The use of the Request-URI, `To`, `From`, and `Call-ID` headers in a REGISTER request is slightly different than for other requests. The Request-URI contains only the

Table 4.1

Mandatory Headers in an `INVITE` `Request`

Call-ID
CSeq
From
To
Via
Contact
Max-Forwards

Table 4.2
Types of Registrar Actions and `Contact` Headers

Request Headers	Registrar Action
Contact: * Expires: 0	Cancel all registrations
Contact: sip:galvani@bologna.edu.it; expires=30	Add Contact to current registrations; registration expires in 30 minutes
Contact: sip:galvani@bologna.edu.it Expires: 30	Add Contact to current registrations; registration expires in 30 minutes
Contact: sip:galvani@bolognauni.edu; expires=45	Add all Contacts to registrations in preference order listed; first one expires in 45 minutes, second in 30 minutes
Contact: sip:l.galvani@bologna.it Expires: 30	
Contact: sip:galvani@bologna.edu.it; action=proxy ;q=0.9	Add Contacts to current registrations using specified preference SIP requests
Contact:mailto:galvani@bologna.edu.it; q=0.1	should be proxied; SIP URI expires in 60 minutes (default) mailto URL does not expire
No Contact header present	Return all current registrations in response

domain of the registrar server with no user portion. The REGISTER request may be forwarded or proxied until it reaches the authoritative registrar server for the specified domain. The To header contains the SIP URI of the AOR of the user agent that is being registered. The From contains the SIP URI of the sender of the request, usually the same as the To header. It is recommended that the same Call-ID be used for all registrations by a user agent.

A user agent sending a REGISTER request may receive a 3xx redirection or 4xx failure response containing a Contact header of the location to which registrations should be sent.

A third-party registration occurs when the party sending the registration request is not the party that is being registered. In this case, the From header will contain the URI of the party submitting the registration on behalf of the party identified in the To header. Chapter 3 contains an example of a first-party registration. An example third-party registration request for the user Euclid is shown below:

```
REGISTER sip:registrar.athens.gr SIP/2.0
Via: SIP/2.0/UDP 201.202.203.204:5060;branch=z9hG4bK313
Max-Forwards:70
To: sip:euclid@athens.gr
From: <sip:secretary@academy.athens.gr>;tag=543131
Call-ID: 2000-July-07-23:59:59.1234@201.202.203.204
CSeq: 1 REGISTER
Contact: sip:euclid@parthenon.athens.gr
Contact: mailto:euclid@geometry.org
Content-Length: 0
```

The mandatory headers in a REGISTER request are shown in Table 4.3.

4.1.3   BYE

The BYE method is used to terminate an established media session. In telephony, it is similar to a release message. A session is considered established if an INVITE has received a success class response (2xx) or an ACK has been sent. A BYE is sent only by user agents participating in the session, never by proxies or other third parties. It is an end-to-end method, so responses are only generated by the other user agent. A user agent responds with a 481 Dialog/Transaction Does Not Exist to a BYE for an unknown dialog.

It is not recommended that a BYE be used to cancel pending INVITEs because it will not be forked like an INVITE and may not reach the same set of user agents as the INVITE. An example BYE request looks like the following:

```
BYE sip:info@hypotenuse.org SIP/2.0
Via: SIP/2.0/UDP port443.hotmail.com:5060;branch=z9hG4bK312bc
Max-Forwards:70
To: <sip:info@hypotenuse.org>;tag=63104
From: <sip:pythag42@hotmail.com>;tag=9341123
Call-ID: 34283291273@port443.hotmail.com
```

Table 4.3
Mandatory Headers in a REGISTER Request

| Call-ID |
| CSeq |
| From |
| To |
| Via |
| Max-Forwards |

```
CSeq: 47 BYE
Content-Length: 0
```

The mandatory headers in a BYE request are shown in Table 4.4.

4.1.4 ACK

The ACK method is used to acknowledge final responses to INVITE requests. Final responses to all other requests are never acknowledged. Final responses are defined as 2xx, 3xx, 4xx, 5xx, or 6xx class responses. The CSeq number is never incremented for an ACK, but the CSeq method is changed to ACK. This is so that a UAS can match the CSeq number of the ACK with the number of the corresponding INVITE.

An ACK may contain an application/sdp message body. This is permitted if the initial INVITE did not contain a SDP message body. If the INVITE contained a message body, the ACK may not contain a message body. The ACK may not be used to modify a media description that has already been sent in the initial INVITE; a re-INVITE must be used for this purpose. SDP in an ACK is used in some interworking scenarios with other protocols where the media characteristics may not be known when the initial INVITE is generated and sent. An example of this is described in Section 10.6.

For 2xx responses, the ACK is end-to-end, but for all other final responses it is done on a hop-by-hop basis when stateful proxies are involved. The end-to-end nature of ACKs to 2xx responses allows a message body to be transported. An ACK generated in a hop-by-hop acknowledgment will contain just a single Via header with the address of the proxy server generating the ACK. The difference between hop-by-hop acknowledgments to a response end-to-end acknowledgments is shown in the message fragments of Figure 4.2.

A hop-by-hop ACK reuses the same branch ID as the INVITE since it is considered part of the same transaction. An end-to-end ACK uses a different branch ID as it is considered a new transaction.

Table 4.4
Mandatory Headers in a BYE Request

Call-ID
CSeq
From
To
Via
Max-Forwards

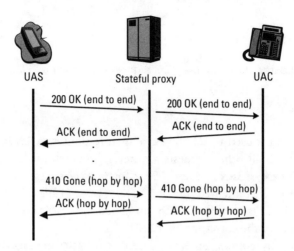

Figure 4.2 End-to-end versus hop-by-hop acknowledgments.

A stateful proxy receiving an ACK message must determine whether or not the ACK should be forwarded downstream to another proxy or user agent or not. That is, is the ACK a hop-by-hop ACK or an end-to-end ACK? This is done by comparing the branch ID for a match pending transaction branch IDs. If there is not an exact match, the ACK is proxied toward the UAS. Otherwise, the ACK is for this hop and is not forwarded by the proxy. The call flows of Chapter 10 show examples of both types of ACK handling. An example ACK containing SDP contains:

```
ACK sip:laplace@mathematica.org SIP/2.0
Via: SIP/2.0/TCP 128.5.2.1:5060;branch=z9hG4bK1834
Max-Forwards:70
To: Marquis de Laplace <sip:laplace@mathematica.org>;tag=90210
From: Nathaniel Bowditch <sip:n.bowditch@salem.ma.us>;tag=887865
Call-ID: 152-45-N-32-23-W@128.5.2.1
CSeq: 3 ACK
Content-Type: application/sdp
Content-Length: 143

v=0
o=bowditch 2590844326 2590944532 IN IP4 salem.ma.us
s=Bearing
c=IN IP4 salem.ma.us
t=0 0
m=audio 32852 RTP/AVP 0
a=rtpmap:0 PCMU/8000
```

The mandatory and optional headers in an ACK message are shown in Table 4.5.

Table 4.5
Mandatory Headers in an ACK Request

```
Call-ID
CSeq
From
To
Via
Max-Forwards
```

4.1.5 CANCEL

The CANCEL method is used to terminate pending searches or call attempts. It can be generated by either user agents or proxy servers provided that a 1xx response containing a tag has been received, but no final response has been received. A user agent uses the method to cancel a pending call attempt it had earlier initiated. A forking proxy can use the method to cancel pending parallel branches after a successful response has been proxied back to the UAC. CANCEL is a hop-by-hop request and receives a response generated by the next stateful element. The difference between a hop-by-hop request and an end-to-end request is shown in Figure 4.3. The CSeq is not incremented for this method so that proxies and user agents can match the CSeq of the CANCEL with the CSeq of the pending INVITE to which it corresponds.

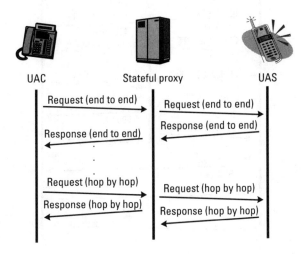

Figure 4.3 End-to-end versus hop-by-hop requests.

The branch ID for a CANCEL matches the INVITE that it is canceling. A CANCEL only has meaning for an INVITE since only an INVITE may take several seconds (or minutes) to complete. All other SIP requests complete immediately (that is, a UAS must immediately generate a final response). Consequently, the final result will always be generated before the CANCEL is received.

A proxy receiving a CANCEL forwards the CANCEL to the same set of locations with pending requests that the initial INVITE was sent to. A proxy does not wait for responses to the forwarded CANCEL requests, but responds immediately. A user agent confirms the cancellation with a 200 OK response to the CANCEL and replies to the INVITE with a 487 Request Terminated response.

If a final response has already been received, a user agent will need to send a BYE to terminate the session. This is also the case in the race condition where a CANCEL and a final response cross in the network, as shown in Figure 4.4. In this example, the CANCEL and 200 OK response messages cross between the proxy and the UAS. The proxy still replies to the CANCEL with a 200 OK, but then also forwards the 200 OK response to the INVITE. The 200 OK response to the CANCEL sent by the proxy only means that the CANCEL request was received and has been forwarded—the UAC must still be prepared

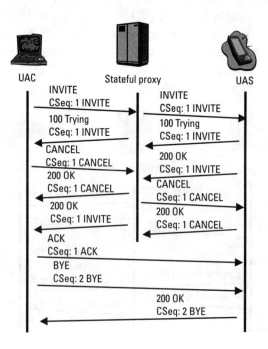

Figure 4.4 Race condition in call cancellation.

to receive further final responses. No 487 response is sent in this scenario. The session is canceled by the UAC sending an ACK then a BYE in response to the 200 OK.

Since it is a hop-by-hop request, a CANCEL may not contain a message body. An example CANCEL request contains:

```
CANCEL sip:i.newton@cambridge.edu.gb SIP/2.0
Via: SIP/2.0/UDP 10.downing.gb:5060
 ;branch=z9hG4bK3134134
Max-Forwards:70
To: Isaac Newton <sip:i.newton@cambridge.edu.gb>
From: Rene Descartes <sip:visitor@10.downing.gb>;tag=034323
Call-ID: 42@10.downing.gb
CSeq: 32156 CANCEL
Content-Length: 0
```

The mandatory header fields in a CANCEL request are shown in Table 4.6.

4.1.6 OPTIONS

The OPTIONS method is used to query a user agent or server about its capabilities and discover its current availability. The response to the request lists the capabilities of the user agent or server. A proxy never generates an OPTIONS request. A user agent or server responds to the request as it would to an INVITE (i.e., if it is not accepting calls, it would respond with a 4xx or 6xx response). A success class (2xx) response can contain Allow, Accept, Accept-Encoding, Accept-Language, and Supported headers indicating its capabilities.

An OPTIONS request may not contain a message body. A proxy determines if an OPTIONS request is for itself by examining the Request-URI. If the Request-URI contains the address of the proxy, the request is for the proxy. Otherwise, the OPTIONS is for another proxy or user agent and the request is forwarded. An example OPTIONS request and response contains:

Table 4.6
Mandatory Headers in a CANCEL Request

Call-ID
CSeq
From
To
Via
Max-Forwards

```
OPTIONS sip:user@carrier.com SIP/2.0
Via: SIP/2.0/UDP cavendish.kings.cambridge.edu.uk
 ;branch=z9hG4bK1834
Max-Forwards:70
To: <sip:user@proxy.carrier.com>
From: J.C. Maxwell <sip:james.maxwell@kings.cambridge.edu.uk>
 ;tag=34
Call-ID: 9352812@cavendish.kings.cambridge.edu.uk
CSeq: 1 OPTIONS
Content-Length: 0

SIP/2.0 200 OK
Via: SIP/2.0/UDP cavendish.kings.cambridge.edu.uk;tag=512A6
 ;branch=z9hG4bK0834 ;received=192.0.0.2
To: <sip:user@proxy.carrier.com>;tag=432
From: J.C. Maxwell <sip:james.maxwell@kings.cambridge.edu.uk>
 ;tag=34
Call-ID: 9352812@cavendish.kings.cambridge.edu.uk
CSeq: 1 OPTIONS
Allow: INVITE, OPTIONS, ACK, BYE, CANCEL, REFER
Accept-Language: en, de, fr
Content-Length: ...
Content-Type: application/sdp

v=0
etc...
```

The mandatory headers in an OPTIONS request is the same as Table 4.3.

4.1.7 REFER

The REFER method [2] is used by a user agent to request another user agent to access a URI or URL resource. The resource is identified by a URI or URL in the required Refer-To header field (see Section 6.2.16). Note that the URI or URL can be any type of URI: sip, sips, http, pres and so forth. When the URI is a sip or sips URI, the REFER is probably being used to implement a call transfer service. REFER can also used to implement peer-to-peer call control.

A REFER request can be sent either inside or outside an existing dialog. A typical call flow is shown in Figure 4.5. In this example, a UAC sends a REFER to a UAS. The UAS, after performing whatever authentication and authorization, decides to accept the REFER and responds with a 202 Accepted response. Note that this response is sent immediately without waiting for the triggered request to complete. This is important because REFER uses the non-INVITE method state machine, which requires an immediate final response, unlike an INVITE which may take several seconds (or even minutes) to complete. Since the Refer-To URI in this example is a sip URI, the UAC sends an INVITE setting the Request-URI to the Refer-To URI. This

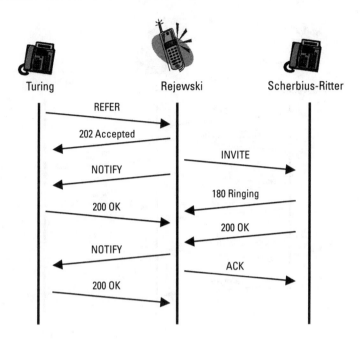

Turing   Rejewski   Scherbius-Ritter

Figure 4.5 REFER example call flow.

INVITE is successful since it receives a 200 OK response. This successful out-
come is communicated back to the UAC using a NOTIFY method (described in
Section 4.1.9). The message body of the NOTIFY contains a partial copy of the
final response to the triggered request. In this case, it contains the start-line
SIP/2.0 200 OK. This part of a SIP message is described in the
Content-Type header field as a message/sipfrag [3].

An example of a REFER message is shown below:

```
REFER sip:m.rejewski@biuroszyfrow.pl SIP/2.0
Via SIP/2.0/UDP lab34.bletchleypark.co.uk:5060
 ;branch=z9hG4bK932039
Max-Forwards: 69
To: <sip:m.rejewski@biuroszyfrow.pl>;tag=ACEBDC
From: Alan Turing <sip:turing@bletchleypark.co.uk>;tag=213424
Call-ID: 3419fak3kFD23s1A9dkl
CSeq: 5412 REFER
Refer-To: <sip:info@scherbius-ritter.com>
Content-Length: 0
```

Another example is the use of REFER to "push" a Web page. In the example of Figure 4.6, the UAC sends a REFER to the UAS with a Refer-To set
to an HTTP URL or a Web page. This causes the UAS to send a 202
Accepted then send a HTTP GET request to the Web server identified by the

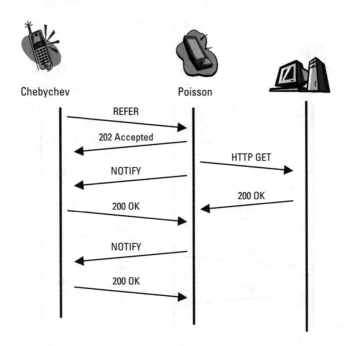

Figure 4.6 REFER example used to push Web page.

URL. After the Web page has loaded, the UAS sends a NOTIFY containing a body and HTTP/1.0 200 OK.

A REFER and the SIP request triggered by the REFER may contain the Referred-By header field (see Section 6.2.17), which contains information about who requested the request.

Figure 4.7 shows an advanced use of REFER to implement a common PSTN or PBX feature known as attended transfer [4]. In this feature, the transferor is assumed to be in a dialog (in a session) with the transferee. The transferor places the transferee on hold, then sends an INVITE to another party, called the transfer target. After the session is established between the transferor and the transfer target, the transferor then puts the target on hold. Now the transferor has two on hold sessions. The transferor then sends a REFER to the transferee which causes the transferee to generate a new INVITE (called a "triggered" INVITE) to the target. The successful INVITE "replaces" the existing session between the transferor and the transfer target. When the transferee receives notification that the transfer was successful, the session between the transferor and the transferee is terminated with a BYE. This application uses "escaped" header fields in the Refer-To URI. That is, certain SIP header fields are specified and prepopulated in the URI, which are then copied into the

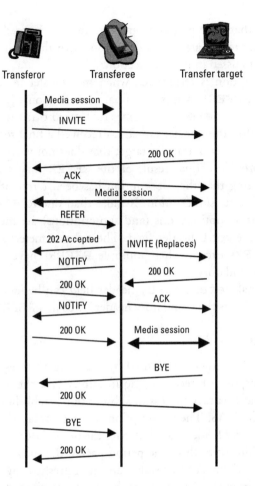

Figure 4.7 Use of REFER and Replaces to perform attended transfer feature.

triggered INVITE. In this case, the transferor generates the Replaces (see Section 6.2.23) header field necessary in the triggered INVITE to make the transfer succeed. The transferee copies the escaped Replaces header and places it in the INVITE.

The acceptance of a REFER with a 202 Accepted response creates an implicit subscription (a subscription without sending a SUBSCRIBE request; see Section 4.1.8). After sending the 202 Accepted, the target must send an immediate NOTIFY with the status 100 Trying and Subscription-State: active;expires=60, which indicates that the subscription will expire in 60 seconds (the expiration value is chosen by the notifier). The Subscription-State header contains the expiration time of the subscription. If that time period expires before the triggered request has completed, both sides terminate the subscription, with the notifier sending a final notification as discussed in the next section.

The subscription is terminated when the transfer target (the party that accepted the REFER) sends a final notification (a NOTIFY with Subscription-State: terminated;reason=noresource). Usually, this is after the transfer target has received a final response to the triggered request. However, a transfer target that does not wish to establish a subscription and provide a final result of the REFER may send an immediate NOTIFY indicating that the subscription has been terminated. Each REFER sent creates a separate subscription. If more than one REFER is sent within a dialog, the resulting notifications (and subscriptions) are identified by an id parameter in the Event header field. The id parameter is optional in refer triggered NOTIFYs except when multiple REFERs have been accepted, in which case it is mandatory.

The optional Referred-By header field can be included in a REFER request. Table 4.7 lists the mandatory header fields in a REFER request.

4.1.8   SUBSCRIBE

The SUBSCRIBE method [5] is used by a user agent to establish a subscription for the purpose of receiving notifications (via the NOTIFY method) about a particular event. A successful subscription establishes a dialog between the UAC and the UAS. The subscription request contains an Expires (see Section 6.4.7) header field, which indicates the desired duration of the existence of the subscription. After this time period passes, the subscription is automatically terminated. The subscription can be refreshed by sending another SUBSCRIBE within the dialog before the expiration time. A server accepting a subscription returns a 200 OK response also containing an Expires header

Table 4.7
Mandatory Header Fields for a REFER

To
From
Call-ID
CSeq
Contact
Max-Forwards
Via
Refer-To

field. The expiration timer can be the same as the request, or the server may shorten the interval, but it may not lengthen the interval. There is no "UNSUBSCRIBE" method used in SIP—instead a SUBSCRIBE with Expires:0 requests the termination of a subscription and hence the dialog. A terminated subscription (either due to timeout out or a termination request) will result in a final NOTIFY indicating that the subscription has been terminated (see Section 4.1.9 on NOTIFY).

A 202 Accepted response to a SUBSCRIBE does not indicate whether the subscription has been authorized—it merely means it has been understood by the server.

The basic call flow is shown in Figure 4.8. The client sends a SUBSCRIBE, which is successful, and receives NOTIFYs as the requested events occur at the server. Before the expiration of the subscription time, the client re-SUBSCRIBEs to extend the subscription and hence receives more notifications.

Note that a client must be prepared to receive a NOTIFY before receiving a 200 OK response to the SUBSCRIBE. Also, due to forking, a client must be prepared to receive NOTIFYs from multiple servers (the NOTIFYs will have

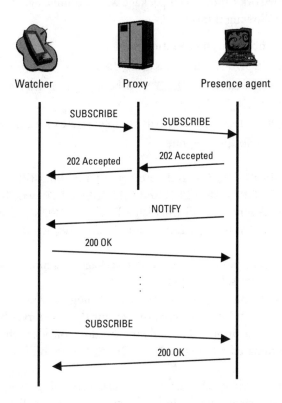

Figure 4.8 Example SUBSCRIBE and NOTIFY call flow.

different To tags and hence will establish separate dialogs), although only one
200 OK response to the SUBSCRIBE may be received.

An example SUBSCRIBE request is shown below:

```
SUBSCRIBE sip:ptolemy@rosettastone.org SIP/2.0
Via SIP/2.0/UDP proxy.elasticity.co.uk:5060
 ;branch=z9hG4bK348471123
Via SIP/2.0/UDP parlour.elasticity.co.uk:5060
 ;branch=z9hG4bKABDA ;received=192.0.3.4
Max-Forwards: 69
To: <sip:Ptolemy@rosettastone.org>
From: Thomas Young <sip:tyoung@elasticity.co.uk>;tag=1814
Call-ID: 452k59252058dkfj349241k34
CSeq: 3412 SUBSCRIBE
Allow-Events: dialog
Contact: <sip:tyoung@parlour.elasticity.co.uk>
Event: dialog
Content-Length: 0
```

The type of event subscription is indicated by the required Event header
field (see Section 6.4.7) in the SUBSCRIBE request. Each application of the
SIP Events framework [5] defines a package with a unique event tag. Each package defines the following things:

- Default subscription expiration interval;
- Expected SUBSCRIBE message bodies;
- What events cause a NOTIFY to be sent, and what message body is
  expected in the NOTIFY;
- Whether the NOTIFY contains complete state or increments (deltas);
- Maximum notification rate.

A protocol called PSTN and Internet Interworking (PINT) [6] defined
methods SUBSCRIBE, NOTIFY, and UNSUBSCRIBE, which have a similar
semantic to SIP. A server can distinguish a PINT SUBSCRIBE request from a
SIP SUBSCRIBE by the absence of an Event header field in the PINT
request.

A server should indicate which event packages it supports by listing them
in an Allow-Events header field.

If a SUBSCRIBE refresh is sent within a dialog but receives a 481 Dialog Does Not Exist response, this means that the server has already terminated the subscription. The client should consider the dialog and
subscription terminated and send a SUBSCRIBE to establish a new dialog and
subscription.

An event template package is a special type of package that can be
applied to any other package including statistical, access policy, and subscription lists. The application of a template package to a package is shown by

separating the package and template package names with a "." as in `pres-ence.winfo`, which is the application of the watcher info template package to the presence package. Table 4.8 lists the current set of SIP event and template packages.

Table 4.9 lists the manatory header fields in a `SUBSCRIBE` request. Packages are standardized in the SIPPING or SIMPLE WG based on the requirements in [5].

4.1.9 NOTIFY

The `NOTIFY` method [5] is used by a user agent to convey information about the occurrence of a particular event. A `NOTIFY` is always sent within a dialog

Table 4.8
Event Packages and Template Packages

`conference`	Conference information including participant lists, policy information, and so forth [7]
`dialog`	Dialog state and identification information [8]
`message-summary`	Messages notification, used for message waiting indicator (mwi) with voicemail [9]
`presence`	Presence information [10]
`refer`	Refer state implicit subscription created by REFER [2]
`reg`	User registration state [11]
`winfo`	Watcher information template package [12]

Table 4.9
Mandatory Header Fields for a `SUBSCRIBE`

`To`
`From`
`Call-ID`
`CSeq`
`Max-Forwards`
`Via`
`Contact`
`Event`
`Allow-Events`

when a subscription exists between the subscriber and the notifier. However, it is possible for a subscription to be established using non-SIP means (no SUBSCRIBE is sent) and may also be implicit in another SIP request type (for example, a REFER establishes an implicit subscription). Since it is sent within a dialog, the NOTIFY will contain a To tag, From tag, and existing Call-ID. A basic call flow showing NOTIFY is shown in Figure 4.8.

A NOTIFY request normally receives a 200 OK response to indicate that it has been received. If a 481 Dialog/Transaction Does Not Exist response is received, the subscription is automatically terminated and no more NOTIFYs are sent.

NOTIFY requests contain an Event header field indicating the package and a Subscription-State header field indicating the current state of the subscription. The Event header field (see Section 6.2.6) will contain the package name used in the subscription. Currently defined packages are listed in Table 4.8 The Subscription-State header field (see Section 6.2.27) will either be active, pending, or terminated.

A NOTIFY is always sent at the start of a subscription and at the termination of a subscription. If a NOTIFY contains incremental (delta) state information, the message body will contain a state version number that will be incremented by 1 for each NOTIFY sent. This way, the receiver of the NOTIFY can tell if information is missing our received out of sequence.

An example NOTIFY request is shown here:

```
NOTIFY sip:tyoung@parlour.elasticity.co.uk SIP/2.0
Via SIP/2.0/UDP cartouche.rosettastone.org:5060
 ;branch=z9hG4bK3841323
Max-Forwards: 70
To: Thomas Young <sip:tyoung@elasticity.co.uk>;tag=1814
From: <sip:ptolemy@rosettastone.org>;tag=5363956k
Call-ID: 452k59252058dkfj349241k34
CSeq: 3 NOTIFY
Contact: <sip:Ptolemy@cartouche.rosettastone.org>
Event: dialog
Subscription-State: active
Allow-Events: dialog
Content-Type: application/xml+dialog
Content-Length: ...

(XML Message body not shown...)
```

Table 4.10 lists the mandatory header fields in a NOTIFY request.

4.1.10 MESSAGE

The MESSAGE method [13] is used to transport instant messages (IM) using SIP. IM usually consists of short message exchanged in near-real time by

Table 4.10
Mandatory Header Fields for a NOTIFY

```
To
From
Call-ID
CSeq
Max-Forwards
Via
Contact
Event
Subscription-State
Allow-Events
```

participants engaged in a "conversation." MESSAGEs may be sent within a dialog or outside a dialog, but they do not establish a dialog by themselves. The actual message content is carried in the message body as a MIME attachment. All UAs that support the MESSAGE method must support plain/text format, they may support other formats such as message/cpim [14] or text/html, or many others.

A MESSAGE request normally receives a 200 OK response to indicate that the message has been delivered to the final destination. An Instant Message response should not be sent in the message body of a 200 OK, but rather a separate MESSAGE request sent to the original sender. A 202 Accepted response indicates that the request has reached a store-and-forward device and will likely eventually be delivered to the final destination. In neither case does the 2xx response confirm that the message content has been rendered to the user.

A MESSAGE request may use the im (Instant Message) URI scheme [14] in a Request-URI, although a client should try to resolve to a sip or sips.

An example MESSAGE call flow is shown in Figure 4.9.

Note that the MESSAGE method is not the only application of instant messaging with SIP. It is also possible to use SIP to establish an instant message session in a completely analogous way that SIP is commonly used to establish a media session. An INVITE could be used to establish the session with a SDP body that describes the instant message protocol to be used directly between the two users. IM sessions have been proposed using SIP MESSAGE, Common Presence and Instant Messaging (CPIM) [14], and even Jabber [15]. An example call flow is shown in Figure 4.10.

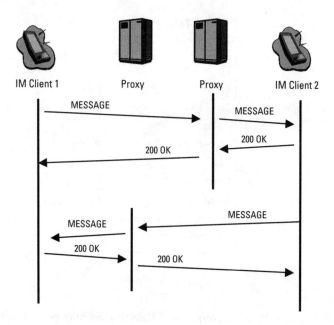

Figure 4.9 SIP instant message example.

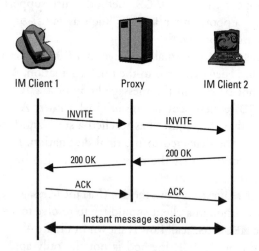

Figure 4.10 Using SIP to establish an instant messaging session.

An example MESSAGE request is shown below:

```
MESSAGE sip:editor@rcs.org SIP/2.0
Via SIP/2.0/UDP lab.mendeleev.org:5060;branch=z9hG4bK3
Max-Forwards: 70
```

```
To: <editor@rcs.org>
From: "D. I. Mendeleev" <dmitry@mendeleev.org>;tag=1865
Call-ID: 93847197172049343
CSeq: 5634 MESSAGE
Subject: First Row
Contact: <sip:dmitry@lab.mendeleev.org>
Content-Type: text/plain
Content-Length: 5

H, He
```

Table 4.11 lists the mandatory header fields in a MESSAGE request.

4.1.11 INFO

The INFO [16] method is used by a user agent to send call signaling information to another user agent with which it has an established media session. This is different from a re-INVITE since it does not change the media characteristics of the call. The request is end-to-end, and is never initiated by proxies. A proxy will always forward an INFO request—it is up to the UAS to check to see if the dialog is valid. INFO requests for unknown dialogs receive a 481 Transaction/Dialog Does Not Exist response.

An INFO method typically contains a message body. The contents may be signaling information, a midcall event, or some sort of stimulus. INFO has been proposed to carry certain PSTN midcall signaling information such as ISUP USR messages.

The INFO method always increments the CSeq. An example INFO method is:

```
INFO sip:poynting@mason.edu.uk SIP/2.0
Via: SIP/2.0/UDP cavendish.kings.cambridge.edu.uk
 ;branch=z9hG4bK24555
Max-Forwards: 70
```

Table 4.11
Mandatory Header Fields for a MESSAGE

To
From
Call-ID
CSeq
Max-Forwards
Via

```
To: John Poynting <sip:nting@mason.edu.uk> ;tag=3432
From: J.C. Maxwell <sip:james.maxwell@kings.cambridge.edu.uk>
 ;tag=432485820183
Call-ID: 18437@cavendish.kings.cambridge.edu.uk
CSeq: 6 INFO
Content-Type: message/isup
Content-Length: 16

51a6324134527
```

The mandatory headers in an INFO request are shown in Table 4.12.

4.1.12 PRACK

The PRACK [17] method is used to acknowledge receipt of reliably transported provisional responses (1xx). The reliability of 2xx, 3xx, 4xx, 5xx, and 6xx responses to INVITEs is achieved using the ACK method. However, in cases where a provisional response, such as 180 Ringing, is critical in determining the call state, it may be necessary for the receipt of a provisional response to be confirmed. The PRACK method applies to all provisional responses except the 100 Trying response, which is never reliably transported.

A PRACK is generated by a UAC when a provisional response has been received containing a RSeq reliable sequence number (see Section 6.3.10) and a Supported: 100rel header. The PRACK echoes the number in the RSeq and the CSeq of the response in a RAck header. The message flow is as shown in Figure 4.5. In this example, the UAC sends the 180 Ringing response reliably by including the RSeq header. When no PRACK is received from the UAC after the expiration of a timer, the response is retransmitted. The receipt of the PRACK confirms the delivery of the response and stops all further transmissions. The 200 OK response to the PRACK stops retransmissions of the PRACK request. The call completes when the UAC sends the ACK in response to the 200 OK.

Table 4.12
Mandatory Headers in an INFO Request

Mandatory Headers
Call-ID
CSeq
From
To
Via
Max-Forwards

Reliable responses are retransmitted using the same exponential backoff mechanism used for final responses to an INVITE. The combination of Call-ID, CSeq number, and RAck number allows the UAC to match the PRACK to the provisional response it is acknowledging. As shown in Figure 4.5, the PRACK receives a 200 OK response, which can be distinguished from the 200 OK to the INVITE by the method contained in the CSeq header. The detailed use of the method (see Figure 4.11) is described in Sections 6.2.29 and 6.3.10 (where the RAck and RSeq headers are described).

The PRACK method always increments the CSeq. A PRACK may contain a message body. An example exchange contains:

```
SIP/2.0 180 Ringing
Via: SIP/2.0/UDP lucasian.trinity.cambridge.edu.uk
 ;branch=z9hG4bK452352
 ;received=1.2.3.4
To: Descartes <sip:rene.descartes@metaphysics.org>;tag=12323
From: Newton <sip:newton@kings.cambridge.edu.uk>;tag=981
```

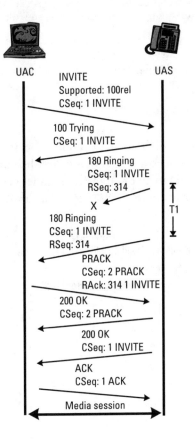

Figure 4.11 Use of reliable provisional responses.

```
Call-ID: 5@lucasian.trinity.cambridge.edu.uk
RSeq: 314
CSeq: 1 INVITE
Content-Length: 0

PRACK sip:rene.descartes@metaphysics.org SIP/2.0
Via: SIP/2.0/UDP lucasian.trinity.cambridge.edu.uk
 ;branch=z9hG4bKdtyw
Max-Forwards: 70
To: Descartes <sip:rene.descartes@metaphysics.org>;tag=12323
From: Newton <sip:newton@kings.cambridge.edu.uk>;tag=981
Call-ID: 5@lucasian.trinity.cambridge.edu.uk
CSeq: 2 PRACK
RAck: 314 1 INVITE
Content-Length: 0

SIP/2.0 200 OK
Via: SIP/2.0/UDP lucasian.trinity.cambridge.edu.uk
 ;branch=z9hG4bKdtyw ;received=1.2.3.4
To: Descartes <sip:rene.descartes@metaphysics.org>;tag=12323
From: Newton <sip:newton@kings.cambridge.edu.uk>;tag=981
Call-ID: 5@lucasian.trinity.cambridge.edu.uk
CSeq: 2 PRACK
Content-Length: 0
```

The mandatory header fields in a PRACK request are shown in Table 4.13.

4.1.13 UPDATE

The UPDATE method [18] is used to modify the state of a session without changing the state of the dialog. A session is established in SIP using an INVITE request (see Section 4.1.1) in an offer/answer manner (see Section 7.1.12). Typically, a session offer is made in the INVITE and an answer made in a response to

Table 4.13
Mandatory Header Fields in a PRACK Request

Mandatory Header Fields
Call-ID
CSeq
From
To
Via
Max-Forwards
RAck

the INVITE. In an established session, a re-INVITE is used to update session parameters. However, neither party in a pending session (INVITE sent but no final response received) may re-INVITE—instead, the UPDATE method is used.

Possible uses of UPDATE include muting or placing on hold pending media streams, performing QoS or other end-to-end attribute negotiation prior to session establishment

The following is an example UPDATE message (Figure 4.12):

```
UPDATE sips:beale@bufords.bedford.va.us SIP/2.0
Via SIP/2.0/TLS client.crypto.org:5060;branch=z9hG4bK342
Max-Forwards: 70
To: T. Beale <sips:beale@bufords.bedford.va.us>;tag=71
From: Blaise Vigenere <sips:bvigenere@crypto.org>;tag=19438
Call-ID: 170189761183162948
CSeq: 94 UPDATE
Contact: <sips:client.crypto.org>
Content-Type: application/sdp
Content-Length: ...

(SDP Message body not shown...)
```

Table 4.14 lists the mandatory header fields in an UPDATE request.

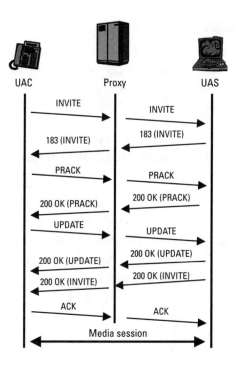

Figure 4.12 Example of UPDATE method.

Table 4.14
Mandatory Header Fields for an UPDATE

```
To
From
Call-ID
Cseq
Max-Forwards
Via
Contact
```

4.2   URI and URL Schemes Used by SIP

SIP supports a number of URI and URL schemes including sip, sips, tel, pres, and im for SIP, secure SIP, telephone, presence, and instant message URIs as described in the following sections. In addition, other URI schemes can be present in SIP header fields as listed in Table 4.15.

4.2.1   SIP and SIPS URIs

The addressing scheme of SIP URLs and URIs has been previously mentioned. SIP URIs are used in a number of places including the To, From, and Contact headers, as well as the Request-URI, which indicates the destination. SIP URIs are similar to the mailto URL [20] and can be used in hyperlinks

Table 4.15
Common URI Schemes Used in SIP Messages

Scheme	Meaning	Use
sip	SIP URI	Most common in To, From, Contact, Request-URIs, and so forth [1]
sips	Secure SIP URI	Same as SIP, provides end-to-end encryption using TLS [1]
tel	Telephone URI	Represents a telephone number (E.164), usually in To or From [19]
pres	Presence URI	Used to represent the URI of a presence agent [14]
im	Instant Message URI	Instant Message client [14]
mailto	E-mail URL	Can be included in Contact in registration or redirection response [20]
http	Web URL	Can be used in some headers such as Alert-Info, Error-Info, and so forth [21]

on Web pages, for example. SIP URIs can include telephone numbers. The information in a SIP URI indicates the way in which the resource (user) should be contacted using SIP.

An example SIP URI contains the scheme `sip` a ":", then a `user-name@host` or IPv4 address followed by an optional ":", then the port number, or a list of ";" separated by URI parameters:

```
sip:joseph.fourier@transform.org:5060;transport=udp;user=ip
;method=INVITE;ttl=1;maddr=240.101.102.103?Subject=FFT
```

Some SIP URIs, such as a `REGISTER` Request-URI do not have a user-name, but begin with the host or IPv4 address. In this example, the port number is shown as `5060`, the well-known port number for SIP. For a SIP URI, if the port number is not present, `5060` is assumed. For a SIPS URI, port number `5061` is assumed. The `transport` parameter indicates UDP is to be used, which is the default. TCP, TLS, and SCTP are alternative transport parameters.

The `user` parameter is used by parsers to determine if a telephone number is present in the username portion of the URI. The assumed default is that it is not, indicated by the value `ip`. If a telephone number is present, it is indicated by the value `phone`. This parameter must not be used to guess at the characteristics or capabilities of the user agent. For example, the presence of a `user=phone` parameter must not be interpreted that the user agent is a SIP telephone (which may have limited display/processing capabilities). In a telephony environment, IP telephones and IP/PSTN gateways may in fact use the reverse assumption, interpreting any digits in a username as digits regardless if `user=phone` is present.

The `method` parameter is used to indicate the method to be used. The default is `INVITE`. This parameter has no meaning in `To` or `From` header fields or in a Request-URI but can be used in `Contact` headers for registration, for example, or in a `Refer-To` header field.

The `ttl` parameter is the time-to-live, which must only be used if the maddr parameter contains a multicast address and the transport parameter contains `udp`. The default value is 1. This value scopes the multicast session broadcast, as described in Section 1.8.

The `maddr` usually contains the multicast address to which the request should be directed, overriding the address in the host portion of the URI. It can also contain, however, a unicast address of an alternative server for requests.

The `method`, `maddr`, `ttl`, and `header` parameters must not be included in `To` or `From` headers, but may be included in `Contact` headers or in Request-URIs. In addition to these parameters, a SIP URI may contain other user-defined parameters.

Following the "?" parameter, names can be specified to be included in the request. This is similar to the operation of the `mailto` URL, which allows

Subject and Priority to be set for the request. Additional headers can be specified, separated by a "&". The header name body indicates that the contents of a message body for an INVITE request is being specified in the URI.

If the parameter user=phone is present, then the username portion of the URI can be interpreted as a telephone number. This allows additional parameters in the username portion of the URI. This allows the parameters and structure of a tel URL [19] to be present in the user part of the SIP URI as described in the next section.

The sips URI scheme has the same structure as the sip URI but begins with the sips scheme name. Note that a sips URI is not equivalent to a sip URI with transport=tls, since the sip URI does not have the same security requirements as the sips URI. The requirement is that TLS transport is used end-to-end for the SIP path. The only exception is hop between the final proxy and the UAS, which may use another security mechanism besides TLS (IPSec, for example).

Not shown in the example is the loose route parameter lr, which can be present in sip or sips Record-Route and Route URIs to indicate that the proxy server identified by the URI supports loose routing.

4.2.2 Telephone URLs

The telephone URI scheme, tel, [19] can be used to represent a resource identified by a telephone number. Telephone numbers can be of two general forms, local or global. A local number is only valid in a particular geographic area and has only local significance. If the number is used outside of this area, it will either fail or return the wrong resource. A global telephone number, also called an E.164 number, is one that is, in principle, valid anywhere. It contains enough information about the country, region, and locality so that the PSTN network is capable of routing calls to the correct resource. An example of a local phone number is:

```
tel:411;phone-context=+1314
```

which indicates a call to directory assistance valid only within country code 1 and area code 314 as identified in the required phone-context parameter. An example of a global phone number is:

```
tel:+13145551212
```

Global phone numbers always begin with the "+" identifier followed by the country code, 1 in this case, followed by the remaining telephone digits.

A tel URL can also contain some characters and information about dialing strings and patterns. For example:

```
tel:#70555-1212;isub=1000
```

In this example, the dialed digit string, interpreted by a PSTN gateway, would be the DTMF digit # then 70 (to cancel call waiting, for example), then the digits 555–1212. Additional parameters include an ISDN subaddress of 1000. This example shows both types of optional visual separators allowed, either "-" or "." as the separator.

Tel URLs can also be embedded in Web pages and could be included in HTML as, for example:

```
Click <A HREF="tel:+1972.555.1212">here</A> to get information
    about Dallas.
```

The syntax and parameters of the `tel` URL may be used in the user portion of a `sip` URI. For example, the first `tel` example could be represented as a `sip` URI as follows:

```
sip:411%3Bphone-context%3D+1314@gateway.example.com
```

The SIP URI adds a domain portion which represents the domain or gateway that will route the request. Note that some of the characters in the `tel` URI must be escaped if they are included in the user part of a `sip` URI, in this example the ";" and "=" are escaped.

4.2.3 Presence and Instant Messaging URLs

The `pres` URL scheme is defined [14] as a URL scheme that represents a "presentity" or presence agent. The `im` URL scheme is defined [14] as a URL scheme that represents an "instant inbox" or an instant message client. Both URL schemes do not represent a new protocol but are resolved using DNS SRV resource records, which return another URI that indicates the actual presence or instant messaging protocol. For example, if the presence agent reference by the presence URL:

```
pres:user@example.com
```

supports SIP presence, the DNS SRV query would return a SIP URI, for example:

```
sip:user@example.com
```

which would then allow a presence agent to send a SUBSCRIBE to this SIP URI to obtain the presence agent of this user.

The same procedure would be used for resolving an `im` URL into a SIP URI for sending a MESSAGE request.

4.3 Tags

A tag is a cryptographically random number with at least 32 bits of randomness, which is added to To and From headers to uniquely identify a dialog. The examples of Chapters 2 and 10 show the use of the tag header parameter. The To header in the initial INVITE will not contain a tag. A caller must include a tag in the From header, although a RFC 2543 user agent generally will not do so as it was optional in that specification. Excluding 100 Trying, all responses will have a tag added to the To header. The sending or reception of a response containing a From tag creates an early dialog. A tag returned in a 200 OK response is then incorporated as a dialog identifier and used in all future requests for this Call-ID. A tag is never copied across calls. Any response generated by a proxy will have a tag added by the proxy. An ACK generated by either a user agent or a proxy will always copy the From tag of the response in the ACK request.

If a UAC receives responses containing different tags, this means that the responses are from different UASs, and hence the INVITE has been forked. It is up to the UAC as to how to deal with this situation. For example, the UAC could establish separate sessions with each of the responding UAS. The dialogs would contain the same From, Call-ID, and CSeq, but would have different tags in the To header. The UAC also could BYE certain legs and establish only one session.

Note that tags are not part of the To or From URI but are part of the header and always placed outside any "<>".

4.4 Message Bodies

Message bodies in SIP may contain various types of information. They may contain SDP information, which can be used to convey media information or QoS or even security information.

The optional Content-Disposition header is used to indicate the intended use of the message body. If not present, the function is assumed to be session, which means that the body describes a media session. Besides session, the other defined function is render, which means that the message body should be presented to the user or otherwise used or displayed. This could be used to pass a small JPEG image file or URI.

The format of the message body is indicated by the Content-Type header described in Section 6.4.5. If a message contains a message body, the message must include a Content-Type header. All user agents must support a Content-Type of application/sdp. The encoding scheme of the message body is indicated in the Content-Encoding header. If not specified, the encoding is assumed to be `text/plain`. The specification of a `Content-Encoding` scheme allows the message body to be compressed.

The `Content-Length` header contains the number of octets in the message body. If there is no message body, the `Content-Length` header should still be included but has a value of 0. Because multiple SIP messages can be sent in a TCP stream, the `Content-Length` count is a reliable way to detect when one message ends and another begins. If a `Content-Length` is not present, the UAC must assume that the message body continues until the end of the UDP datagram, or until the TCP connection is closed, depending on the transport protocol.

Message bodies can have multiple parts if they are encoded using Multi-part Internet Mail Extensions (MIME) [22]. Message bodies in SIP, however, should be small enough so that they do not exceed the UDP MTU of the network. Proxies may reject requests with large message bodies with a `413 Request Entity Too Large` response, since processing large messages can load a server.

As mentioned in the previous section, SIP carries message bodies the same way that e-mails carry attachments. It is possible to carry multiple message bodies within a single SIP message. This is done using a multipart MIME body. The `Content-Type` is listed as `multipart/mime`, and a separator is defined, which is used by the parser to separate the message. Any SIP request or response that can contain a message body may carry a multipart MIME body. An example is in SIP-T (see Section 7.5) in which an `INVITE` carries both a SDP message body (`application/sdp`) and an encapsulated ISUP message (`application/isup`). An example multipart MIME is:

```
INVITE sip:refertarget@carol.example.com SIP/2.0
Via: SIP/2.0/UDP referree.example;branch=z9hG4bKffe209934aac
To: sip:refertarget@carol.example.com
From: <sip:referree@referree.example>;tag=2909034023
Call-ID: fe9023940-a3465@referree.example
CSeq: 889823409 INVITE
Max-Forwards: 70
Contact: <sip:referree@bob.example.com>
Referred-By: sip:referror@alice.example.com
 ;cid=%3C20398823.2UWQFN309shb3@alice.example.com%3E
Content-Type: multipart/mixed; boundary=-*-boundary-*-
Content-Length: ...

--*-boundary-*-

Content-Type: application/sdp
Content-Length: ...

v=0
o=referree 2890844526 2890844526 IN IP4 referree.example
s=Session SDP
```

```
c=IN IP4 referree.example
t=0 0
m=audio 49172 RTP/AVP 0
a=rtpmap:0 PCMU/8000

--*-boundary-*-

Content-Type: multipart/signed;
 protocol="application/pkcs7-signature";
 micalg=sha1; boundary=dragons39
Content-ID: <20398823.2UWQFN309shb3@alice.example.com>
Content-Length: (appropriate value)

-another-boundary-
Content-Type: message/sipfrag
Content-Disposition: auth-id; handling=optional

From: sip:referror@alice.example.com
Date: Thu, 21 Feb 2002 13:02:03 GMT
Call-ID: 2203900ef0299349d9209f023a
Refer-To: sip:refertarget@carol.example.com
Referred-By: sip:referror@alice.example.com
 ;cid=%3C20398823.2UWQFN309shb3@alice.example.com%3E

-another-boundary-
Content-Type: application/pkcs7-signature; name=smime.p7s
Content-Transfer-Encoding: base64
Content-Disposition: attachment; filename=smime.p7s;han-
dling=required

(S/MIME data goes here)

—unique44—
--*-boundary-*--
```

References

[1]   Rosenberg, J., et al., "SIP: Session Initiation Protocol," RFC 3261, 2002.

[2]   Sparks, R., "The Session Initiation Protocol (SIP) Refer Method," RFC 3515, 2003.

[3]   Sparks, R., "Internet Media Type message/sipfrag," RFC 3420, 2003.

[4]   Sparks, R., and A. Johnston, "SIP Call Control – Transfer," IETF Internet-Draft, Work in Progress, March 2003.

[5]   Roach, A., "Session Initiation Protocol (SIP)-Specific Event Notification," RFC 3265, 2002.

[6]   Petrack, S., and L. Conroy, "The PINT Service Protocol: Extensions to SIP and SDP for IP Access to Telephone Call Services," RFC 2848, 2000.

[7]   Rosenberg, J., and H. Schulzrinne, "A Session Initiation Protocol (SIP) Event Package for Conference State," IETF Internet-Draft, Work in Progress, June 2002.

[8]   Rosenberg, J., and H. Schulzrinne, "An INVITE Initiated Dialog Event Package for the Session Initiation Protocol (SIP)," IETF Internet-Draft, Work in Progress, March 2003.

[9]   Mahy, R., "A Message Summary and Message Waiting Indication Event Package for the Session Initiation Protocol (SIP)," IETF Internet-Draft, Work in progress, March 2003.

[10] Rosenberg, J., "A Presence Event Package for the Session Initiation Protocol (SIP)," IETF Internet-Draft, Work in Progress, January 2003.

[11] Rosenberg, J., "A Session Initiation Protocol (SIP) Event Package for Registrations," IETF Internet-Draft, October 2002.

[12] Rosenberg, J., "A Watcher Information Event Template-Package for the Session Initiation Protocol (SIP)," IETF Internet-Draft, Work in Progress, January 2003.

[13] Campbell, B., et al., "Session Initiation Protocol (SIP) Extension for Instant Messaging," RFC 3428, 2002.

[14] Crocker, D., et al., "Common Presence and Instant Messaging (CPIM)," IETF Internet-Draft, Work in Progress, August 2002.

[15] Sparks, R., "Establishing jabber Messaging Sessions with the Session Initiation Protocol," IETF Internet-Draft, Work in Progress, October 2002.

[16] Donovan, S., "The SIP INFO Method," RFC 2976, 2000.

[17] Rosenberg, J., and H. Schulzrinne, "Reliability of Provisional Responses in Session Initiation Protocol (SIP)," RFC 3262, 2002.

[18] Rosenberg, J., "The Session Initiation Protocol (SIP) UPDATE Method," RFC 3311, 2002.

[19] Schulzrinne, H., and A. Vaha-Sipila, "The tel URI for Telephone Calls," IETF Internet-Draft, February 2003.

[20] Hoffman, P., L. Masinter, and J. Zawinski, "The mailto URL Scheme," RFC 2368, 1998.

[21] Fielding, R., et al., "Hypertext Transfer Protocol — HTTP/1.1," RFC 2616, 1999.

[22] Freed, N., and N. Borenstein, "Multipurpose Internet Mail Extensions (MIME) Part One: Format of Internet Message Bodies," RFC 2045, 1996.

5

SIP Response Messages

This chapter covers the types of SIP response messages. A SIP response is a message generated by a UAS or a SIP server to reply to a request generated by a UAC. A response may contain additional header fields containing information needed by the UAC. Or, it may be a simple acknowledgment to prevent retransmissions of the request by the UAC. Many responses direct the UAC to take specific additional steps. The responses are discussed in terms of structure and classes. Then, each request type is discussed and examined in detail.

There are six classes of SIP responses. The first five classes were borrowed from HTTP; the sixth was created for SIP. The classes are shown in Table 5.1.

If a particular SIP response code is not understood by a UAC, it must be interpreted by the class of the response. For example, an unknown 599 Server Unplugged response must be interpreted by a user agent as a 500 Server Failure response.

The reason phrase is for human consumption only—the SIP protocol uses only the response code in determining behavior. Thus, a 200 Call Failed is interpreted the same as 200 OK. The reason phrases listed here are the suggested ones from the RFC document. They can be used to convey more information, especially in failure class responses—the phrase is likely to be displayed to the user. Some response codes were borrowed from HTTP, perhaps with a slightly different reason phrase.[1] However, not all HTTP response codes are valid in SIP, and some even have a different meaning. New response codes

---

1. Not all HTTP response codes are supported in SIP. Only the response codes described in RFC 3261 and supporting RFCs are supported in SIP.

Table 5.1

SIP Response Classes

Class	Description	Action
1xx	Informational	Indicates status of call prior to completion. If first informational or provisional response.
2xx	Success	Request has succeeded. If for an INVITE, ACK should be sent; otherwise, stop retransmissions of request.
3xx	Redirection	Server has returned possible locations. The client should retry request at another server.
4xx	Client error	The request has failed due to an error by the client. The client may retry the request if reformulated according to response.
5xx	Server failure	The request has failed due to an error by the server. The request may be retried at another server.
6xx	Global failure	The request has failed. The request should not be tried again at this or other servers.

created for SIP typically start at x80 to try to avoid collisions with HTTP response codes.

Unless otherwise referenced, the responses described here are defined in RFC 3261 [1].

5.1 Informational

The informational class of responses 1xx are used to indicate call progress. Informational responses are end-to-end responses and may contain message bodies. The exception to this is the 100 Trying response, which is only a hop-by-hop response and may not contain a message body. Any number of informational responses can be sent by a UAS prior to a final response (2xx, 3xx, 4xx, 5xx, or 6xx class response) being sent. The first informational response received by the UAC confirms receipt of the INVITE, and stops retransmission of the INVITE, as described in Section 3.5. For this reason, servers returning 100 Trying responses minimizes INVITE retransmissions in the network. Further informational responses have no effect on INVITE retransmissions. A stateful proxy receiving a retransmission of an INVITE will resend the last provisional response sent to date. Informational responses are optional—a UAS can send a final response without first sending an informational response. While final responses to an INVITE receive an ACK to confirm receipt, provisional responses are not acknowledged, except using the PRACK method described in Section 4.1.12.

All provisional responses with the exception of 100 Trying must contain a Contact URI and echo all Record-Route headers received in the request. However, a RFC 2543 implementation will not do this, as it was not mandated in that document.

5.1.1 100 Trying

This special case response is only a hop-by-hop request. It is never forwarded and may not contain a message body. A forking proxy must send a 100 Trying response, since the extended search being performed may take a significant amount of time. This response can be generated by either a proxy server or a user agent. It only indicates that some kind of action is being taken to process the call—it does not indicate that the user has been located. A 100 Trying response typically does not contain a To tag.

5.1.2 180 Ringing

This response is used to indicate that the INVITE has been received by the user agent and that alerting is taken place. This response is important in interworking with telephony protocols, and it is typically mapped to messages such as an ISDN Progress or ISUP Address Complete Message (ACM) [2]. When the user agent answers immediately, a 200 OK is sent without a 180 Ringing; this scenario is called the "fast answer" case in telephony.

A message body in this response could be used to carry QoS or security information, or to convey ring tone or animations from the UAS to the UAC.

A UA normally generates its own ringback tone or remote ringing indication, unless a Alert-Info header field (see Section 6.1.1) is present.

5.1.3 181 Call Is Being Forwarded

This response is used to indicate that the call has been handed off to another end-point. This response is sent when this information may be of use to the caller. Also, because a forwarding operation may take longer for the call to be answered, this response gives a status for the caller.

5.1.4 182 Call Queued

This response is used to indicate that the INVITE has been received, and will be processed in a queue. The reason phrase can be used to indicate the estimated wait time or the number of callers in line, as shown in Figure 5.1. A message body in this response can be used to carry music on hold or other media.

Caller   Call center

Figure 5.1 Call queuing example with call processing center.

5.1.5   183 Session Progress

The 183 Session Progress response indicates that information about the progress of the session (call state) may be present in a message body or media stream. Unlike a 100 Trying response, a 183 is an end-to-end response and does establish a dialog (must contain a To tag and Contact). Unlike a 180, 181, or 182 response, however, it does not convey any specific information about the status of the INVITE. A typical use of this response is to allow a UAC to hear ring tone, busy tone, or a recorded announcement in calls through a gateway into the PSTN. This is because call progress information is carried in the media stream in the PSTN. A one-way media connection or trunk is established from the calling party's telephone switch to the called party's telephone switch in the PSTN prior to the call being answered. In SIP, the media session is established after the call is answered—after a 200 OK and ACK have been exchanged between the UAC and UAS. If a gateway used a 180 Ringing response instead, no media path would be established between the UAC and the gateway, and the caller would never hear ring tone, busy tone, or a recorded announcement (e.g., "The number you have dialed has changed, the new number is ...") since these are all heard in the media path prior to the call being answered. Figure 5.2 shows an example where a SIP caller does not hear a recorded announcement coming from the PSTN. Figure 5.3 shows the use of the 183 Session Progress allowing an early media session to be established prior to the call being answered. The PSTN interworking scenarios in Chapter 10 show this in detail.

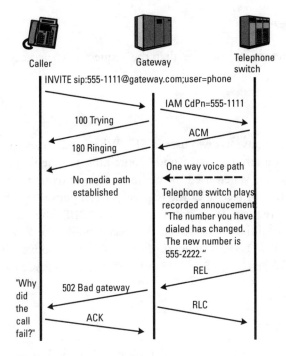

Figure 5.2 PSTN interworking without early media.

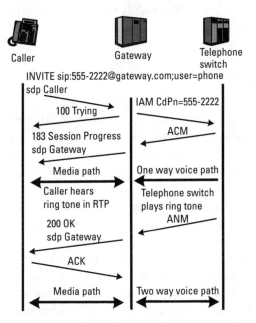

Figure 5.3 PSTN interworking with early media.

5.2 Success

Success class responses indicate that the request has succeeded or has been accepted.

5.2.1 200 OK

The 200 OK response has two uses in SIP. When used to accept a session invitation, it will contain a message body containing the media properties of the UAS (called party). When used in response to other requests, it indicates successful completion or receipt of the request. The response stops further retransmissions of the request. In response to an OPTIONS, the message body may contain the capabilities of the server. A message body may also be present in a response to a REGISTER request. For 200 OK responses to CANCEL, INFO, MESSAGE, SUBSCRIBE, NOTIFY, and PRACK, a message body is not permitted.

5.2.2 202 Accepted

The 202 Accepted response [3] indicates that the UAS has received and understood the request, but that the request may not have been authorized or processed by the server. It is commonly used in responses to SUBSCRIBE (see Section 4.1.8) and REFER (see Section 4.1.7), and sometimes MESSAGE (see Section 4.1.10) methods.

5.3 Redirection

Redirection class responses are generally sent by a SIP server acting as a redirect server in response to an INVITE, as described in Section 3.3.2. A UAS, however, can also send a redirection class response to implement certain types of call forwarding features. There is no requirement that a UAC receiving a redirection response must retry the request to the specified address. The UAC can be configured to automatically generate a new INVITE upon receipt of a redirection class response without requiring user intervention. In addition, proxies may also automatically send an ACK to a redirect and proxy the INVITE to the new location provided in the Contact URI of the redirection. To prevent looping, the server must not return any addresses contained in the request Via header field, and the client must check the address returned in the Contact header field against all other addresses tried in an earlier call attempt. Note that this type of transaction looping is different from request looping.

5.3.1 300 Multiple Choices

This redirection response contains multiple `Contact` header fields, which indicate that the location service has returned multiple possible locations for the `sip` or `sips` URI in the Request-URI. The order of the `Contact` header fields is assumed to be significant. That is, they should be tried in the order in which they were listed in the response.

5.3.2 301 Moved Permanently

This redirection response contains a `Contact` header field with the new permanent URI of the called party. The address can be saved and used in future `INVITE` requests.

5.3.3 302 Moved Temporarily

This redirection response contains a URI that is currently valid but that is not permanent. As a result, the `Contact` header field should not be cached across calls unless an `Expires` header field is present, in which case the location is valid for the duration of the time specified.

5.3.4 305 Use Proxy

This redirection response contains a URI that points to a proxy server who has authoritative information about the calling party. The caller should resend the request to the proxy for forwarding. This response could be sent by a UAS that is using a proxy for incoming call screening. Because the proxy makes the decisions for the UAS on acceptance of the call, the UAS will only respond to `INVITE` requests that come from the screening proxy. Any `INVITE` request received directly would automatically receive this response without user intervention.

5.3.5 380 Alternative Service

This response returns a URI that indicates the type of service that the called party would like. An example might be a redirect to a voicemail server.

5.4 Client Error

This class of response is used by a server or UAS to indicate that the request cannot be fulfilled as it was submitted. The specific client error response or the presence of certain header fields should indicate to the UAC the nature of the error and how the request can be reformulated. The UAC should not resubmit the request without modifying the request based on the response. The same request, however, can be tried in other locations. A forking proxy receipt of a 4xx response does not terminate the search. Typically, client error responses will require user intervention before a new request can be generated.

5.4.1   400 Bad Request

This response indicates that the request was not understood by the server. An example might be a request that is missing required header fields such as To, From, Call-ID, or CSeq. This response is also used if a UAS receives multiple INVITE requests (not retransmissions) for the same Call-ID.

5.4.2   401 Unauthorized

This response indicates that the request requires the user to perform authentication. This response is generally sent by a user agent, since the 407 Proxy Authentication Required (Section 5.4.8) is sent by a proxy that requires authentication. The exception is a registrar server, which sends a 401 Unauthorized response to a REGISTER message that does not contain the proper credentials. An example of this response is:

```
SIP/2.0 401 Unathorized
Via: SIP/2.0/UDP proxy.globe.org:5060;branch=z9hG4bK2311ff5d.1
 ;received=192.0.2.1
Via: SIP/2.0/UDP 173.23.43.1:5060;branch=z9hG4bK4545
From: <sip:explorer@geographic.org>;tag=341323
To: <sip:printer@maps-r-us.com>;tag=19424103
From: Copernicus <sip:copernicus@globe.org>;tag=34kdilsp3
Call-ID: 123456787@173.23.43.1
CSeq: 1 INVITE
WWW-Authenticate: Digest realm="globe.org",
 nonce="8eff88df84f1cec4341ae6e5a359", qop="auth",
 opaque="", stale=FALSE, algorithm=MD5
Content-Length: 0
```

The presence of the required WWW-Authenticate header field is required to give the calling user agent a chance to respond with the correct credentials. A typical authentication exchange using SIP digest is shown in Figure 3.7. Note that the follow-up INVITE request should use the same Call-ID as the original request as the authentication may fail in some cases if the Call-ID is changed from the initial request to the retried request with the proper credentials.

5.4.3   402 Payment Required

This response is a placeholder for future definition in the SIP protocol. It could be used to negotiate call completion charges.

5.4.4 403 Forbidden

This response is used to deny a request without giving the caller any recourse. It is sent when the server has understood the request, found the request to be correctly formulated, but will not service the request. This response is not used when authorization is required.

5.4.5 404 Not Found

This response indicates that the user identified by the `sip` or `sips` URI in the Request-URI cannot be located by the server, or that the user is not currently signed on with the user agent.

5.4.6 405 Method Not Allowed

This response indicates that the server or user agent has received and understood a request but is not willing to fulfill the request. An example might be a `REGISTER` request sent to a user agent. An `Allow` header field (Section 6.4.1) must be present to inform the UAC what methods are acceptable. This is different from the case of an unknown method, in which a `501 Not Implemented` response is returned. Note that a proxy will forward request types it does not understand unless the request is targeted to the proxy server (i.e., the Request-URI is the URI of the proxy server).

5.4.7 406 Not Acceptable

This response indicates that the request cannot be processed due to a requirement in the request message. The `Accept` header field in the request did not contain any options supported by the UAS.

5.4.8 407 Proxy Authentication Required

This request sent by a proxy indicates that the UAC must first authenticate itself with the proxy before the request can be processed. The response should contain information about the type of credentials required by the proxy in a `Proxy-Authenticate` header field. The request can be resubmitted with the proper credentials in a `Proxy-Authorization` header field. Unlike in HTTP, this response may not be used by a proxy to authenticate another proxy.

```
SIP/2.0 407 Proxy Authorization Required
Via: SIP/2.0/UDP discrete.sampling.org:5060;branch=z9hG4bK6563
 ;received=65.64.140.198
From: Shannon <sip:shannon@sampling.org>;tag=59204
To: Schockley <sip:shockley@transistor.com>;tag=142334
Call-ID: adf8gasdd7fld@discrete.sampling.org
```

```
CSeq: 1 INVITE
Proxy-Authenticate: Digest realm="sampling.org", qop="auth",
 nonce="9c8e88df84df1cec4341ae6cbe5a359",
 opaque="", stale=FALSE, algorithm=MD5
Content-Length: 0
```

5.4.9   408 Request Timeout

This response is sent when an Expires header field is present in an INVITE request, and the specified time period has passed. This response could be sent by a forking proxy or a user agent. The request can be retried at any time by the UAC, perhaps with a longer time period in the Expires header field or no Expires header field at all. Alternatively, a stateful proxy can send this response after the request transaction times out without receiving a final response.

5.4.10   409 Conflict

This response code has been removed from RFC 3261 but is defined in RFC 2543. It indicates that the request cannot be processed due to a conflict in the request. This response is used by registrars to reject a registration with a conflicting action parameter.

5.4.11   410 Gone

This response is similar to the 404 Not Found response but contains the hint that the requested user will not be available at this location in the future. This response could be used by a service provider when a user cancels their service.

5.4.12   411 Length Required

This response code has been removed from RFC 3261 but is defined in RFC 2543. This response can be used by a proxy to reject a request containing a message body but no Content-Length header field. A proxy that takes a UDP request and forwards it as a TCP request could generate this response, since the use of Content-Length is more critical in TCP requests. However, the response code is not very useful since a proxy can easily calculate the length of a message body in a UDP request (it is until the end of the UDP packet) but cannot with a stream-oriented transport such as TCP. In this case, a missing Content-Length header field would cause the message body to go on indefinitely, which would generate a 513 Message Too Large response instead of a 411 Length Required.

5.4.13 413 Request Entity Too Large

This response can be used by a proxy to reject a request that has a message body that is too large. A proxy suffering congestion could temporarily generate this response to save processing long requests.

5.4.14 414 Request-URI Too Long

This response indicates that the Request-URI in the request was too long and cannot be processed correctly. There is no maximum length defined for a Request-URI in the SIP standard document.

5.4.15 415 Unsupported Media Type

This response sent by a user agent indicates that the media type contained in the INVITE request is not supported. For example, a request for a video conference to a PSTN gateway that only handles telephone calls will result in this response. The response should contain header fields to help the UAC reformulate the request.

5.4.16 416 Unsupported URI Scheme

The 416 Unsupported URI Scheme response is new to RFC 3261 and is used when a UAC uses a URI scheme in a Request-URI that the UAS does not understand. For example, if a request URI contains a secure SIP (sips) scheme that a proxy does not understand, it would return a 416 response. Since all SIP elements must understand the sip scheme, the request should be retried using a sip uri in the request-uri.

5.4.17 420 Bad Extension

This response indicates that the extension specified in the Require header field is not supported by the proxy or user agent. The response should contain a Supported header field (Section 6.1.12) listing the extensions that are supported. The UAC could resubmit the same request without the extension in the Require header field (Section 6.2.22) or submit the request to another proxy or user agent.

5.4.18 421 Extension Required

The 421 Extension Required response indicates that a server requires an extension to process the request that was not present in a Supported header field in the request. The required extension should be listed in a Required header field in the response. The client should retry the request adding the extension to a `Supported` header field, or try the request at a different server that may not require the extension.

5.4.19   422 Session Timer Interval Too Small

The `422 Session Timer Interval Too Small` response [4] is used to reject a request containing a `Session-Expires` header field (Section 6.2.26) with too short an interval. The ability to reject short durations is important to prevent excessive re-`INVITE` or `UPDATE` traffic. The minimum allowed interval is indicated in the required `Min-SE` header field (Section 6.3.4). The requestor may retry the request without the `Session-Expires` header field or with a value less than or equal to the specified minimum.

5.4.20   423 Interval Too Brief

The `423 Interval Too Brief` response is returned by a registrar that is rejecting a registration request because the requested expiration time on one or more `Contacts` is too brief. The response must contain a `Min-Expires` header field (see Section 6.3.3) listing the minimum expiration interval that the registrar will accept. A client requesting a too short interval can unnecessarily load a registrar server with registration refresh requests—this response allows a registrar to protect against this.

5.4.21   428 Use Authentication Token

The `428 Use Authentication Token` response [5] is used by a UAS that is requiring the use of an Authentication Information Body (AIB) [6]. The AIB is a S/MIME body that is an encrypted `message/sip` or `message/sipfrag` body. At a minimum, an AIB must contain the `From`, `Date`, and `Call-ID` header fields, and they should contain the `To`, `Contact`, and `CSeq` header fields as well. (The selection of these header fields is to prevent cut-and-paste attacks and hijacking.) The SIP request should contain a `Content-Type: message/sipfrag` header field and a `Content-Disposition: aib; handling=optional` header field.

For more information on S/MIME signatures, see Section 3.7.

5.4.22   429 Provide Referror Identity

The `429 Provide Referror Identity` response [7] is used to request that a `Referred-By` header field be re-sent with a valid `Referred-By` security token. The security token is carried as an S/MIME message body. The recipient of this error message (the UA that received and accepted the REFER) should relay this request back to the originator of the REFER by including it in a NOTIFY. The sender of the REFER can then generate the Referred-By security token and include it in the REFER, which would then be copied into the triggered request.

5.4.23   480 Temporarily Unavailable

This response indicates that the request has reached the correct destination, but the called party is not available for some reason. The reason phrase should be modified for this response to give the caller a better understanding of the situation. The response should contain a Retry-After header indicating when the request may be able to be fulfilled. For example, this response could be sent when a telephone has its ringer turned off, or a "do not disturb" button has been pressed. This response can also be sent by a redirect server.

5.4.24   481 Dialog/Transaction Does Not Exist

This response indicates that a response referencing an existing call or transaction has been received for which the server has no records or state information.

5.4.25   482 Loop Detected

This response indicates that the request has been looped and has been routed back to a proxy that previously forwarded the request. Each server that forwards a request adds a Via header with its address to the top of the request. A branch parameter is added to the Via header, which is a hash function of the Request-URI, and the To, From, Call-ID, and CSeq number. A second part is added to the branch parameter if the request is being forked. The reason the branch parameter must be checked is to allow a request to be routed back to a proxy, provided that the Request-URI has changed. This could happen with a call forwarding feature. In this case, the Via headers would differ by having different branch parameters.

5.4.26   483 Too Many Hops

This response indicates that the request has been forwarded the maximum number of times as set by the Max-Forwards header in the request. This is

indicated by the receipt of a Max-Forwards: 0 header in a request. In the
following example, the UAC included a Max-Forwards: 4 header in the
REGISTER request. A proxy receiving this request five hops later generates a
483 response:

```
REGISTER sip:registrar.timbuktu.tu SIP/2.0
Via: SIP/2.0/UDP 201.202.203.204:5060;branch=z9hG4bK45347.1
Via: SIP/2.0/UDP 198.20.2.4:6128;branch=z9hG4bK917a4d4.1
Via: SIP/2.0/UDP 18.56.3.1:5060;branch=z9hG4bK7154.1
Via: SIP/2.0/TCP 101.102.103.104:5060;branch=z9hG4bKa5ff4d3.1
Via: SIP/2.0/UDP 168.4.3.1:5060;branch=z9hG4bK676746
To: sip:explorer@geographic.org
From: <sip:explorer@geographic.org>;tag=341323
Call-ID: 67483010384@168.4.3.1
CSeq: 1 REGISTER
Max-Forwards: 0
Contact: sip:explorer@national.geographic.org
Content-Length: 0

SIP/2.0 483 Too Many Hops
Via: SIP/2.0/UDP 201.202.203.204:5060;branch=z9hG4bK45347.1
Via: SIP/2.0/UDP 198.20.2.4:6128;branch=z9hG4bK917a4d4.1
Via: SIP/2.0/UDP 18.56.3.1:5060;branch=z9hG4bK7154.1
Via: SIP/2.0/TCP 101.102.103.104:5060;branch=z9hG4bKa5ff4d3.1
Via: SIP/2.0/UDP 168.4.3.1:5060;branch=z9hG4bK676746
To: <sip:explorer@geographic.org>;tag=a5642
From: <sip:explorer@geographic.org>;tag=341323
Call-ID: 67483010384@168.4.3.1
CSeq: 1 REGISTER
Content-Length: 0
```

5.4.27   484 Address Incomplete

This response indicates that the Request-URI address is not complete. This
could be used in an overlap dialing scenario in PSTN interworking where digits
are collected and sent until the complete telephone number is assembled by a
gateway and routed [8]. Note that the follow-up INVITE requests may use the
same Call-ID as the original request. An example of overlap dialing is shown
in Figure 5.4.

5.4.28   485 Ambiguous

This request indicates that the Request-URI was ambiguous and must be
clarified in order to be processed. This occurs if the username matches a number
of registrations. If the possible matching choices are returned in Contact
header fields, then this response is similar to the 300 Multiple Choices
response. They are slightly different, however, since the 3xx response returns
equivalent choices for the same user, but the 4xx response returns alternatives
that can be different users. The 3xx response can be processed without human

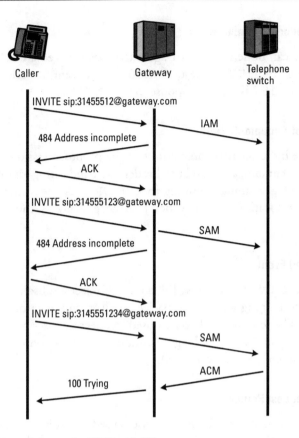

Figure 5.4 Overlap dialing to the PSTN with SIP.

intervention, but this 4xx response requires a choice by the caller, which is why it is classified as a client error class response. A server configured to return this response must take user registration privacy into consideration; otherwise a vague or general Request-URI could be used by a rogue user agent to try to discover sip or sips URIs of registered users.

5.4.29 486 Busy Here

This response is used to indicate that the user agent cannot accept the call at this location. This is different, however, from the 600 Busy Everywhere response, which indicates that the request should not be tried elsewhere. In general, a 486 Busy Here is sent by a UAS unless it knows definitively that the user cannot be contacted. This response is equivalent to the busy tone in the PSTN.

5.4.30   487 Request Terminated

This response can be sent by a user agent that has received a CANCEL request for a pending INVITE request. A 200 OK is sent to acknowledge the CANCEL, and a 487 is sent in response to the INVITE.

5.4.31   488 Not Acceptable Here

This response indicates that some aspect of the proposed session is not acceptable and may contain a Warning header field indicating the exact reason. This response has a similar meaning to 606 Not Acceptable, but only applies to one location and may not be true globally as the 606 response indicates.

5.4.32   489 Bad Event

The 489 Bad Event response [3] is used to reject a subscription request or notification containing an Event package that is unknown or not supported by the UAS. The response code is also used to reject a subscription request that does not specify an Event package, assuming that the server does not support the PINT protocol (see Section 4.1.8).

5.4.33   491 Request Pending

The 491 Request Pending response is used to resolve accidental simultaneous re-INVITEs by both parties in a dialog. Since both INVITEs seek to change the state of the session, they cannot be processed at the same time. While a user agent is awaiting a final response to a re-INVITE, any re-INVITE request received must be replied to with this response code. This is analogous to the "glare" condition in telephony in which both ends seize a trunk at the same time. The reconsideration algorithm in SIP is for the user agent to generate a delay (randomly selected within a range determined by if the user agent send the initial INVITE or not) then retry the re-INVITE, assuming that another re-INVITE has not been received in the meantime. In this way, one side or the other will "win" the race condition and have the re-INVITE processed.

An example is shown in Figure 5.5.

5.4.34   493 Request Undecipherable

The 493 Request Undecipherable response is used when an S/MIME message body can not be decrypted because the public key is unavailable. If the UAS does not support S/MIME, no message body will be present in the response. If the UAS does support S/MIME, the response will contain a message

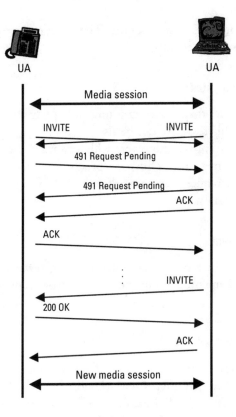

Figure 5.5 Simultaneous re-INVITE resolution example.

body containing a public key suitable for the UAC to use for S/MIME encryption. See Section 3.7 for more details on S/MIME encryption.

5.5 Server Error

This class of responses is used to indicate that the request cannot be processed because of an error with the server. The response may contain a Retry-After header field if the server anticipates being available within a specific time period. The request can be tried at other locations because there are no errors indicated in the request.

5.5.1 500 Server Internal Error

This server error class response indicates that the server has experienced some kind of error that is preventing it from processing the request. The reason phrase

can be used to identify the type of failure. The client can retry the request again
at this server after several seconds.

5.5.2   501 Not Implemented

This response indicates that the server is unable to process the request because it
is not supported. This response can be used to decline a request containing an
unknown method. A proxy, however, will forward a request containing an
unknown request method. Thus, a proxy will forward an unknown
SELF-DESTRUCT request, assuming that the UAS will generate this response
if the method is not known.

5.5.3   502 Bad Gateway

This response is sent by a proxy that is acting as a gateway to another network,
and indicates that some problem in the other network is preventing the request
from being processed.

5.5.4   503 Service Unavailable

This response indicates that the requested service is temporarily unavailable.
The request can be retried after a few seconds, or after the expiration of the
Retry-After header field. Instead of generating this response, a loaded
server may just refuse the connection. This response code is important in that its
receipt triggers a new DNS lookup to locate a backup server to obtain the
desired service. The set of SIP DNS procedures for locating SIP servers is
detailed in [9].

5.5.5   504 Gateway Timeout

This response indicates that the request failed due to a timeout encountered in
the other network to which that the gateway connects. It is a server error class
response because the call is failing due to a failure of the server in accessing
resources outside the SIP network.

5.5.6   505 Version Not Supported

This response indicates that the request has been refused by the server because of
the SIP version number of the request. The detailed semantics of this response
have not yet been defined because there is only one version of SIP (version 2.0)
currently implemented. When additional version numbers are implemented in
the future, the mechanisms for dealing with multiple protocol versions will need
to be detailed.

5.5.7 513 Message Too Large

The 513 Message Too Large response is used by a UAS to indicate that the request size was too large for it to process.

5.6 Global Error

This response class indicates that the server knows that the request will fail wherever it is tried. As a result, the request should not be sent to other locations. Only a server that has definitive knowledge of the user identified by the Request-URI in every possible instance should send a global error class response. Otherwise, a client error class response should be sent. A Retry-After header field can be used to indicate when the request might be successful.

5.6.1 600 Busy Everywhere

This response is the definitive version of the 486 Busy Here client error response. If there is a possibility that the call to the specified Request-URI could be answered in other locations, this response should not be sent.

5.6.2 603 Decline

This response has the same effect as the 600 Busy Everywhere but does not give away any information about the call state of the server. This response could indicate the called party is busy, or simply does not want to accept the call.

5.6.3 604 Does Not Exist Anywhere

This response is similar to the 404 Not Found response but indicates that the user in the Request-URI cannot be found anywhere. This response should only be sent by a server that has access to all information about the user.

5.6.4 606 Not Acceptable

This response can be used to implement some session negotiation capability in SIP. This response indicates that some aspect of the desired session is not acceptable to the UAS, and as a result, the session cannot be established. The response may contain a Warning header field with a numerical code describing exactly what was not acceptable. The request can be retried with different media session information. An example of simple negotiation with SIP is shown in Figure 5.6. If more complicated negotiation capability is required, another protocol should be used.

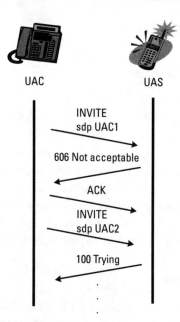

Figure 5.6 Session negotiation with SIP.

References

[1]   Rosenberg, J., et al., "SIP: Session Initiation Protocol," RFC 3261, 2002.

[2]   Camarillo, G., et al., "Integrated Services Digital Network (ISDN) User Part (ISUP) to Session Initiation Protocol (SIP) Mapping," RFC 3398, 2002.

[3]   Roach, A., "SIP Specific Events," RFC 3265, 2002.

[4]   Donovan, S., and J. Rosenberg, "Session Timers in the Session Initiation Protocol (SIP)," IETF Internet-Draft, Work in Progress, November 2002.

[5]   Peterson, J., "Enhancements for Authenticated Identity Management in the Session Initiation Protocol (SIP)," IETF Internet-Draft, Work in Progress, February 2003.

[6]   Peterson, J., "SIP Authenticated Identity Body (AIB) Format," IETF Internet-Draft, Work in Progress, February 2003.

[7]   Sparks, R., "The SIP Referred-By Mechanism," IETF Internet-Draft, Work in Progress, February 2003.

[8]   Anttalainen, T., *Introduction to Telecommunications Network Engineering*, Norwood, MA: Artech House, 1999.

[9]   Rosenberg, J., and H. Schulzrinne, "Session Initiation Protocol (SIP): Locating SIP Servers," RFC 3263, 2002.

6

SIP Header Fields

This chapter describes the header fields present in SIP messages. In RFC 2543, there were four categories of SIP header fields: general, request, response, and entity. RFC 3261 removed this distinction since it was not used by the protocol. However, the header fields discussed in this chapter are categorized as request and response, request only, response only, and message body header fields. Except as noted, header fields are defined in the SIP specification RFC 3261 [1].

SIP header fields in most cases follow the same rules as HTTP header fields [2]. Header fields are defined as `Header: field`, where `Header` is the case-insensitive token (but conventionally lower case with some capitalization) used to represent the header field name, and `field` is the case-insensitive set of tokens that contain the information. Except when otherwise noted, their order in a message is not important. Header fields can continue over multiple lines as long as the line begins with at least one space or horizontal tab character. Unrecognized header fields are ignored by proxies. Many common SIP header fields have a compact form, where the header field name is denoted by a single lower-case character. These header fields are shown in Table 6.1. Header fields can be either end-to-end or hop-by-hop. Hop-by-hop header fields are the only ones that a proxy may insert or, with a few exceptions, modify. Because SIP typically involves end-to-end control, most header fields are end-to-end. The hop-by-hop header fields that may be inserted by a proxy are shown in Table 6.2.

Table 6.1

Compact Forms of SIP Header Fields

Header Field	Compact Form
Accept-Contact	a
Allow-Event	u
Call-ID	i
Contact	m
Content-Encoding	e
Content-Length	l
Content-Type	c
Event	o
From	f
Refer-To	r
Referred-By	b
Reject-Contact	j
Subject	s
To	t
Via	v

6.1   Request and Response Header Fields

This set of header fields can be present in both requests and responses.

6.1.1   Alert-Info

The Alert-Info header field can be used to provide a "distinctive ring" service. If present in an INVITE, the UAS may use the URI to fetch an alert tone to be used in place of the default alerting tone—that is, it would be rendered to the called party. If present in a 180 Ringing response, the UAC may use the URI to fetch a ring-back tone to be rendered to the calling party. In both uses, the URI is fetched and rendered without user intervention, so careful policy rules are necessary to avoid unwanted sounds and noises being generated.

One use is for a trusted proxy to insert the header field with a local (to the domain of the user agent) URI. This then allows for very simple policy in the user agent in deciding whether or not to render.

An example is shown here:

Table 6.2
Header Fields that May Be Inserted or Modified by Proxies

Hop-by-Hop Header Fields
Alert-Info
Call-Info
Content-Length
Date
Error-Info
Max-Forwards
Organization
Priority
Proxy-Authenticate
Proxy-Authorization
Proxy-Require
Record-Route
Reason
Require
Route
Via
WWW-Authenticate

```
Alert-Info: <http://www.provider.com/tones/internal_caller.pcm>
```

6.1.2 Allow-Events

The `Allow-Events` header field [3] is used to list the support event packages that are supported. A UA that supports SIP Events then knows that it may send a `SUBSCRIBE` to that event package. The list of currently defined packages is in Table 4.8. The compact form is `u`.

Examples are shown here:

```
Allow-Events: dialog
u: conference
```

6.1.3 Call-ID

The `Call-ID` header field is mandatory in all SIP requests and responses. It is part of the dialog used to uniquely identify a call between two user agents. A

`Call-ID` must be unique across calls, except in the case of a `Call-ID` in registration requests. All registrations for a user agent should use the same `Call-ID`. A `Call-ID` is always created by a user agent and is never modified by a server.

The `Call-ID` is usually made up of a local-id, which should be a cryptographically random identifier, or a local-id, the @ symbol, and a host name or IP address. Because a user agent can ensure that its local-id is unique within its domain, the addition of the globally unique host name makes the `Call-ID` globally unique. Some security is provided by the randomness of the `Call-ID`, because this prevents a third party from *guessing* a `Call-ID` and presenting false requests. The compact form of the `Call-ID` header field is `i`.

Examples of `Call-ID` are shown here:

```
Call-ID: 34a5d553192cc35@15.34.3.1
Call-ID: 44fer23ei4291dekfer34231232
i: 35866383092031257@port34.carrier.com
```

6.1.4   Contact

The `Contact` header field is used to convey a URI that identifies the resource requested or the request originator, depending on whether it is present in a request or response. Once a `Contact` header field has been received, that URI can be cached and used for routing future requests within a dialog. For example, a `Contact` header field in a `200 OK` response to an `INVITE` can allow the acknowledgment ACK message and all future requests during this call to bypass proxies and go directly to the called party. However, the presence of `Record-Route` header fields in an earlier request or default proxy routing configuration in a user agent may override that behavior. When a `Contact` URI is used in a Request-URI, all URI parameters are allowed with the exception of the `method` parameter, which is ignored.

`Contact` header fields must be present in `INVITE` requests and `200 OK` responses to invitations. In some cases, the `Contact` URI may not resolve directly to the user agent. For example, a UA behind a firewall ALG will need to use a `Contact` URI that resolves to the firewall ALG address. Otherwise, the use of the user agent's URI will result in the call failing because of the firewall blocking any direct routed SIP requests. `Contact` header fields may also be present in `1xx`, `2xx`, `3xx`, and `485` responses. Only in a `REGISTER` request, a special `Contact:*`, along with an `Expires: 0`, header field is used to remove all existing registrations. Examples of `Contact` header fields in registrations are shown in Table 4.3. Otherwise, wild carding is not allowed. A `Contact` header field may contain a display name that can be in quotes. If a display name is present, the URI will be enclosed in < >. If any header field parameters are present, the URI will also be enclosed in < > along with any URI parameters, with the header field parameters outside the < >, even if no display name is present.

There are three additional parameters defined for use in `Contact` header fields: `q`, `action`, and `expires`. They are placed at the end of the URI and separated by semicolons. The `q` value parameter is used to indicate relative preference, which is represented by a decimal number in the range 0 to 1. The `q` value is not a probability, and there is no requirement that the `q` values for a given list of `Contacts` add up to 1. (The `action` parameter defined in RFC 2543 has been deprecated and is not used in RFC 3261. It was only used in registration `Contact` header fields, and is used to specify `proxy` or `redirect` operation by the server.) The `expires` parameter indicates how long the URI is valid and is also only used in registrations. The parameter either contains an integer number of seconds or a date in SIP form (see Section 6.1.4). Examples are shown in Table 6.3.

The `Contact` header field may contain a feature tag [4], which can be used to indicate the capabilities of the device identified by the `Contact` URI. For example, the feature tag `isfocus` is used to indicate that the URI in the `Contact` header field is a Conference URI, and that the dialog is associates with a focus. A focus is a SIP user agent that hosts a particular instance of a conference, called a "bridge" or MCU in other protocols. The presence of the `isfocus` feature tag can be used by a SIP user agent that supports advanced

Table 6.3
Examples of Contact Header Fields

Header Field	Meaning
`Contact: sip:bell@telephone.com`	A single SIP URI without a display name.
`Contact: Lentz <h.lentz@petersburg.edu>`	A display name with the URI is enclosed in < >; the display name is treated as a token and ignored.
`Contact: M. Faraday <faraday@effect.org>, "Faraday" <mailto:faraday@pop.effect.org>`	Two URIs are listed, the second being a non-SIP URI with a display name enclosed in quotes.
`m: <morse@telegraph.org;transport=tcp>; expires= "Fri, 13, Oct 1998 12:00:00 GMT"`	The compact form of the header field is used for a single URI. The URI contains a port number and a URI parameter contained within the < >. An expires header field parameter uses a SIP date enclosed in the quotes.

conferencing features to invoke certain call control operations [5] or subscribe to the conference package [6].

Other feature tags are listed in Table 6.4. The compact form is m.

6.1.5   CSeq

The command sequence CSeq header field is a required header field in every request. The CSeq header field contains a decimal number that increases for each request. Usually, it increases by 1 for each new request, with the exception of CANCEL and ACK requests, which use the CSeq number of the INVITE request to which it refers.

The CSeq count is used by UASs to determine out-of-sequence requests or to differentiate between a new request (different CSeq) or a retransmission (same CSeq). The CSeq header field is used by UACs to match a response to the request it references. For example, a UAC that sends an INVITE request then a CANCEL request can tell by the method in the CSeq of a 200 OK response if it is a response to the invitation or cancellation request. Examples are shown in Table 6.5.

Each user agent maintains its own command sequence number space. For example, consider the case where user agent 1 establishes a session to user agent 2 and initializes its CSeq count to 1. When user agent 2 initiates a request (such as a INVITE or INFO, or even BYE) it will initialize its own CSeq space, totally independent of the CSeq count used by user agent 1. The examples of Chapter 10 show this behavior of CSeq.

6.1.6   Date

The Date header field is used to convey the date when a request or response is sent. The format of a SIP date is based on HTTP dates, but allows only the

Table 6.4
Boolean Feature Tags

Feature tag	Meaning
attendant	Attendant, human or automata
automata	Nonhuman
image	Supports images
message	Supports messaging
text	Supports text
voicemail	Is a voicemail server
isfocus	Is a focus, a conference server

Table 6.5
CSeq Header Field Examples

Header Field	Meaning
`CSeq: 1 INVITE`	The command sequence number has been initialized to 1 for this initial INVITE
`CSeq: 432 REFER`	The command sequence number is set to 432 for this REFER request
`CSeq: 6787 INVITE`	If this was the first request by the user agent for this dialog then either the CSeq was initialized to 6787, or the previous request generated for this Call-ID (either an INVITE or other request) would have had a CSeq of 6786 or lower.

preferred Internet date standard referenced by RFC 1123 [7]. To keep user agent date and time logic simple, SIP only supports the use of the GMT time zone. This allows time entries that are stored in date form rather than second count to be easily converted into delta seconds without requiring knowledge of time zone offsets.

A Date example is shown here:

```
Date: Fri, 13 Oct 1998 23:29:00 GMT
```

6.1.7 Encryption

The Encryption header field was defined in RFC 2543 but is not included in RFC 3261. Instead, encryption using S/MIME is defined as discussed in Section 3.9.

6.1.8 From

The From header field is a required header field that indicates the originator of the request. It is one of two addresses used to identify the dialog. The From header field contains a URI, but it may not contain the transport, maddr, or ttl URI parameters. A From header field may contain a tag, used to identify a particular call. A From header field may contain a display name, in which case the URI is enclosed in < >. If there is both a URI parameter and a tag, then the URI including any parameters must be enclosed in < >. Examples are shown in Table 6.6.

A From tag was optional in RFC 2543 but is mandatory to include in RFC 3261.

Table 6.6

Examples of From Header Field

Header Field	Meaning
From: <sip:armstrong@hetrodyne.com> ;tag=3342436	A single SIP URI with a tag
From: Thomas Edison <sips:edison@electric.com>;tag=532	A secure SIP URI with a display name
f: "James Bardeen" <sip:555.1313@telephone.com ;transport=tcp>;tag=3a320f03	Using the compact form of the header field, a display name in quotes along with a SIP URI with a parameter inside the < >.
From: tel:911	A tel URI without a display name or tag, so no <> is required. Generated by a RFC 2543 UA

6.1.9   Organization

The Organization header field is used to indicate the organization to which the originator of the message belongs. It can also be inserted by proxies as a message is passed from one organization to another. Like all SIP header fields, it can be used by proxies for making routing decisions and by user agents for making call screening decisions.

An example is below:

```
Organization: MCI
```

6.1.10   Record-Route

The Record-Route header field is used to force routing through a proxy for all subsequent requests in a session between two user agents. Normally, the presence of a Contact header field allows user agents to send messages directly bypassing the proxy chain used in the initial request (which probably involved database lookups to locate the called party). A proxy inserting its address into a Record-Route header field overrides this and forces future requests to include a Route header field containing the address of the proxy that forces this proxy to be included.

A proxy, such as a firewall control proxy, wishing to implement this inserts the header field containing its own URI, or adds its URI to an already present Record-Route header field. The URI is constructed so that the URI resolves back to the proxy server. The UAS copies the Record-Route header field into the 200 OK response to the request. The header field is forwarded unchanged by proxies back to the UAC. The UAC then stores the Record-Route proxy list plus a Contact header field if present in the 200 OK for use in a Route header field in all subsequent requests. Because Record-Route is bidirectional, messages in the reverse direction will also traverse the same set of proxies. Chapter 10 contains an example of the use of the Record-Route and Route header fields. The lr parameter is new to RFC 3261 and indicates that the proxy server supports "loose routing." Older RFC 2543 compliant proxy servers create Record-Route URIs that instead of the lr parameter often contain the maddr parameter with an address or host that resolves to that proxy server.

Examples are:

```
Record-Route: <sip:proxy1.carrier.com;lr>,
 <sip:firewall33.corporation.com;lr>

Record-Route:<sip:139.23.1.44;lr>
```

6.1.11 Retry-After

The Retry-After header field is used to indicate when a resource or service may be available again. In 503 Service Unavailable responses, it indicates when the server will be available. In 404 Not Found, 600 Busy Everywhere, and 603 Decline responses, it indicates when the called user agent may be available again.

The header field can also be included by proxy and redirect servers in responses if a recent registration was removed with a Retry-After header field indicating when the user may sign on again. The contents of the header field can be either an integer number of seconds or a SIP date. A duration parameter can be used to indicate how long the resource will be available after the time specified. Examples of this header field are shown in Table 6.7.

6.1.12 Subject

The optional Subject header field is used to indicate the subject of the media session. It can be used by user agents to do simple call screening. The contents of the header field can also be displayed during alerting to aid the user in deciding whether to accept the call. The compact form of this header field is s. Some examples are:

```
Subject: More good info about SIP
s: Are you awake, yet??
```

Table 6.7
Examples of Retry-After Header Field

Header Field	Meaning
`Retry-After: 3600`	Request can be retried again in 1 hour
`Retry-After: Sat, 21 May 2000 08:00:00 GMT`	Request can be retried after the date listed
`Retry-After: 3600`	Request can be tried after 1 hour
`Retry-After: Mon, 29 Feb 2000 13:30:00 GMT ;duration=1800`	Request can be retried after the specified date for 30 minutes

6.1.13 Supported

The `Supported` header field is used to list one or more options implemented by a user agent or server. It is typically included in responses to `OPTIONS` requests. If no options are implemented, the header field is not included. If a UAC lists an option in a `Supported` header field, proxies or UASs may use the option during the call. If the option must be used or supported, the `Require` header field is used instead. Table 6.8 shows the current set of defined feature tags.

An example of the header field is:

    Supported: rel100

6.1.14 Timestamp

The `Timestamp` header field is used by a UAC to mark the exact time a request was generated in some numerical time format. A UAS must echo the

Table 6.8
Extension Feature Tags

Tag	Meaning
`events`	SIP Events [3]
`join`	Join call control primitive [8]
`path`	Path header field [9]
`rel100`	Reliable provisional response (PRACK) support [10]
`replaces`	Replaces call control primitive [11]
`timer`	Session Timer feature [12]

header field in the response to the request and may add another numerical time entry indicating the amount of delay. Unlike the `Date` header field, the time format is not specified. The most accurate time format should be used, including a decimal point. Examples are shown in Table 6.9.

6.1.15 To

The `To` header field is a required header field in every SIP message used to indicate the recipient of the request. Any responses generated by a user agent will contain this header field with the addition of a tag. (Note that an RFC 2543 client will typically only generate a `tag` if more than one `Via` header field is present in the request.) Any response generated by a proxy must have a `tag` added to the `To` header field. A tag added to the header field in a `200 OK` response is used throughout the call and incorporated into the dialog. The `To` header field is never used for routing—the Request-URI is used for this purpose. An optional display name can be present in the header field, in which case the SIP URI is enclosed in < >. If the URI contains any parameters or username parameters, the URI must be enclosed in < > even if no display name is present. The compact form of the header field is `t`. Examples are shown in Table 6.10.

6.1.16 User-Agent

The `User-Agent` header field is used to convey information about the user agent originating the request. Based on the HTTP header field of the same name [2], it can contain manufacturer information, software version, or comments. The field may contain multiple tokens, with the ordering assumed to be from most general to most specific. This information can be used for logging or for generating a specific response for a specific user agent. For security reasons, this header field may be suppressed. For example, an attacker probing a UA for vulnerabilities could learn the particular vendor and software load that is susceptible to a particular attack and reuse that attack against other UAs that have the same software as identified by the `User-Agent` header field.

Examples include:

Table 6.9
Examples of Timestamp Header Field

Header Field	Meaning
Timestamp: 235.15	Client has stamped a start time for the request.
Timestamp: 235.15 .95	This header field from the response has the delay time added by the server.

Table 6.10

Examples of To Header Field

Header Field	Meaning
To: sip:babage@engine.org;tag=2443a8f7	A single SIP URI with a tag and without a display name
To: Thomas Edison <sips:edison@electric.com>	A display name is used, so the sips URI is enclosed in < >
t: "Jim B." <brattain@bell.org>	A display name in quotes along with a SIP URI enclosed within < >
To: <+1-314-555-1212@carrier.com ;user=phone>;tag=8f7f7ad6675	Both a URI parameter and tag are used, so URI is enclosed in < >

```
User-Agent: Acme SIP Phone v2.2
User-Agent: IP Carrier Gateway Av6.4
```

6.1.17 Via

The required Via header field is used to record the SIP route taken by a request and is used to route a response back to the originator. A user agent generating a request records its own address in a Via header field in the request. While the ordering of most SIP header fields is not significant, the Via header fields order is significant because it is used to route responses. A proxy forwarding the request adds a Via header field containing its own address to the *top* of the list of Via header fields. A proxy adding a Via header field always includes a branch tag containing a cryptographic hash of the To, From, Call-ID header fields and the Request-URI. A proxy or user agent generating a response to a request copies all the Via header fields from the request *in order* into the response, then sends the response to the address specified in the top Via header field. A proxy receiving a response checks the top Via header field to ensure that it matches its own address. If it does not, the response has been misrouted and should be discarded. The top Via header field is then removed, and the response forwarded to the address specified in the next Via header field. A simplified Via decision tree is shown in Figure 6.1.

Via header fields contain protocol name and version number and transport (SIP/2.0/UDP, SIP/2.0/TCP, etc.) and may contain port numbers and parameters such as received, branch, maddr, and ttl. A received tag is added to a Via header field if a user agent or proxy receives the request from a different address than that specified in the top Via header field. This indicates that a NAT or firewall proxy is in the message path. If

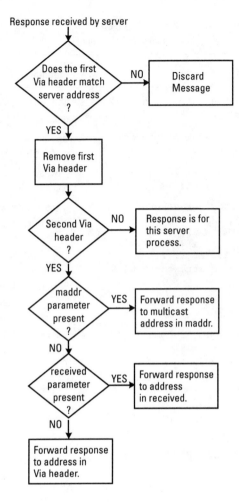

Figure 6.1 Via forwarding decision tree.

present, the received tag is used in response routing. (The hidden parameter, deprecated in RFC 3261, was used to indicate the Via header field has been encrypted.) A branch parameter is added to Via header fields by UAs and proxies, which is computed as a hash function of the Request-URI, and the To, From, Call-ID and CSeq number. A second part is added to the branch parameter if the request is being forked as shown in Figure 3.4. The maddr and ttl parameters are used for multicast transport and have a similar meaning as the equivalent SIP URI parameters. The compact form of the header field is v. Examples are given in Table 6.11.

Table 6.11

Examples of Via Header Field

Header Field	Meaning
Via: SIP/2.0/UDP 100.101.102.103 ;branch=z9hG4bK776a	IPv4 address using unicast UDP transport and assumed port of 5060
Via: SIP/2.0/TCP cube451.office.com:60202 ;branch=z9hG4bK776a	Domain name using TCP transport and port number 60202
Via: SIP/2.0/UDP 120.121.122.123 ;branch= z9hG4bK56a234f3.1	Proxy added Via header field with branch
v: SIP/2.0/UDP proxy.garage.org ;branch= z9hG4bK3423423a3.3	Compact form with domain name using UDP; third search location of forking proxy
Via: SIP/2.0/TCP 192.168.1.2 ;received=12.4.5.50 ;branch=z9hG4bK334	The address is a multicast address. IPv4 address is nonglobally unique. Request has been forwarded through a NAT, which changed the IP address to a globally unique one.
Via:SIP/2.0/UDP host.user.com:4321 ;maddr=224.1.2.3 ;ttl=15 ;branch=z9hG4bK341344	The address is a multicast address specified in maddr with a specified TTL

6.2   Request Header Fields

This set of header fields can only be present in a request.

6.2.1   Accept

The Accept header field is defined by HTTP [2] and is used to indicate acceptable message Internet media types [13] in the message body. The header field describes media types using the format type/sub-type commonly used in the Internet. If not present, the assumed acceptable message body format is application/sdp. A list of media types can have preferences set using q value parameters. The wildcard "*" can be used to specify all sub-types. Examples are given in Table 6.12.

6.2.2   Accept-Contact

The Accept-Contact [4] header field specifies to which URIs the request may be proxied. Some additional parameters are also defined for Contact

Table 6.12
Examples of Accept Header Field

Header Field	Meaning
Accept: application/sdp	This is the default assumed even if no Accept header field is present
Accept: text/*	Accept all text encoding schemes
Accept: application/h.245;q=0.1, application/sdp;q=0.9	Use SDP if possible, otherwise, use H.245

header fields such as media, duplex, and language. This header field is part of the caller preferences extensions to SIP, which have been defined to give some control to the caller in the way a proxy server processes a call. The compact form is a.

Some examples follow:

```
Accept-Contact: *;language=en
a: *;media=video
```

6.2.3 Accept-Encoding

The Accept-Encoding header field, defined in HTTP [2], is used to specify acceptable message body encoding schemes. Encoding can be used to ensure a SIP message with a large message body fits inside a single UDP datagram. The use of q value parameters can set preferences. If none of the listed schemes are acceptable to the UAC, a 406 Not Acceptable response is returned. If not included, the assumed encoding will be text/plain. Examples include:

```
Accept-Encoding: text/plain
Accept-Encoding: gzip
```

6.2.4 Accept-Language

The Accept-Language header field, defined in HTTP [2], is used to specify preferences of language. The languages specified can be used for reason phrases in responses, informational header fields such as Subject, or in message bodies. The HTTP definition allows the language tag to be made of a primary tag and an optional subtag. This header field could also be used by a proxy to route to a human operator in the correct language. The language tags are registered by IANA. The primary tag is an ISO-639 language abbreviation. The use of q values allows multiple preferences to be specified. Examples are shown in Table 6.13.

Table 6.13
Examples of Accept-Language Header Field

Header Field	Meaning
Accept-Language: fr	French is the only acceptable language
Accept-Language: en, ea	Acceptable languages include both English and Spanish
Accept-Language: ea; q=0.5, en ;q=0.9, fr ;q=0.2	Preferred languages are English, Spanish, and French, in that order

6.2.5   Authorization

The Authorization header field is used to carry the credentials of a user agent in a request to a server. It can be sent in reply to a 401 Unauthorized response containing challenge information, or it can be sent first without waiting for the challenge if the form of the challenge is known (e.g., if it has been cached from a previous call). The authentication mechanism for HTTP digest is described in Section 3.8. Examples are shown in Table 6.14.

6.2.6   Call-Info

The Call-Info header field is included in a request by a UAC or proxy to provide a URI with information relating to the session setup. It may be present in an INVITE, OPTIONS or REGISTER request. The header field parameter purpose indicates the purpose of the URI and may have the values icon, info, card, or other IANA registered tokens.

An example follows:

```
Call-Info: <http://www.code.com/my_picture.jpg>;purpose=icon
```

Table 6.14
Example of Authorization Header Field

Header Field	Meaning
Authorization: Digest username="Cust1", realm="company.com", nonce="9c8e88df84f1cec4341ae6e5a359", opaque="", uri="sip:user2@company.com", response="e56131d19580cd833064787ecc"	This HTTP digest authorization header field contains the credentials of Cust1; the nonce was supplied by the SIP server located at the uri specified. The response contains the encrypted username and password. No opaque string is present.

6.2.7 Event

The Event header field is used in a SUBSCRIBE (see Section 4.1.8) or NOTIFY (see Section 4.1.9) methods to indicate which event package is being used by the method. In a SUBSCRIBE, it lists the event package to which the client would like to subscribe. In a NOTIFY, it lists the event package that the notification contains state information about. Currently defined event packages are listed in Table 4.8. The compact form is o.

An example follows:

```
Event: dialog
o: refer
```

6.2.8 Hide

The Hide header field was defined in RFC 2543 but has been deprecated from RFC 3261. It was intended to be used by user agents or proxies to request that the next hop proxy encrypts the Via header fields to hide message routing path information. Encrypted Via headers were identified with the hidden Via parameter. However, the security provided and the mechanism requiring next hop trust made the value of this header field minimal.

6.2.9 In-Reply-To

The In-Reply-To header field is used to indicate the Call-ID that this request references or is returning. For example, a missed call could be returned with a new INVITE and the Call-ID from the missed INVITE copied into the In-Reply-To header field. This allows the UAS to determine that this is not an unsolicited call, which could be used to override call screening logic, for example. Examples of this header field are as follows:

```
In-Reply-To: a8-43-73-ff-43@company.com
In-Reply-To: 12934375@persistance.org, 12934376@persistance.org
```

6.2.10 Join

The Join header field [8] is used in an INVITE to request that the dialog (session) be joined with an existing dialog (session). The parameters of the Join header field identify the dialog by the Call-ID, To tag, and From tag in a similar way to the Replaces header field.

If the Join header field references a point-to-point dialog between two user agents, the Join header field is effectively a request to turn the call into a conference call. If the dialog is already part of a conference, the Join header field is a request to be added into the conference. An example call flow is shown in Figure 6.2 in which a two-way call is turned into a conference call.

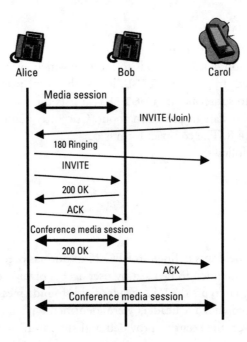

Figure 6.2 Use of Join to create a conference call.

If the dialog referenced in the Join header field does not exist, a 481 Call/Dialog Does Not Exist response is returned. A UA supporting Join should indicate this in all requests with a Supported: join header field.

In the following example, the dialog:

```
To: <sip:moe@example.org>;tag=42312
From: <sip:larry@server.org>;tag=3443212
Call-ID: 243134123234
```

would match the Join header field:

```
Join: 243134123234;to-tag=42312;from-tag=3443212
```

6.2.11 Priority

The Priority header field is used by a UAC to set the urgency of a request. Defined values are non-urgent, normal, urgent, and emergency. This header field could be used to override screening or by servers in load-shedding mechanisms. Because this header field is set by the user agent, it may not be possible for a carrier network to use this field to route emergency traffic, for example. An example is:

```
Priority: emergency
```

6.2.12 Privacy

The Privacy header field [14] is used by a UAC to request varying degrees and types of privacy. Currently defined tags include critical, header, id, session, user, or none.

An example follows:

```
Privacy: header;user;critical
```

6.2.13 Proxy-Authorization

The Proxy-Authorization header field is to carry the credentials of a user agent in a request to a server. It can be sent in reply to a 407 Proxy Authentication Required response containing challenge information, or it can be sent first without waiting for the challenge if the form of the challenge is known (e.g., if it has been cached from a previous call). The authentication mechanism for SIP digest is described in Section 3.9. A proxy receiving a request containing a Proxy-Authorization header field searches for its own realm. If found, it processes the entry. If the credentials are correct, any remaining entries are kept in the request when it is forwarded to the next proxy. An example of this is in Figure 6.3.

Examples are shown in Table 6.15.

6.2.14 Proxy-Require

The Proxy-Require header field is used to list features and extensions that a user agent requires a proxy to support in order to process the request. A 420 Bad Extension response is returned by the proxy listing any unsupported feature in an Unsupported header field. Because proxies by default ignore header fields and features they do not understand, the use of a Proxy-Require header field is needed for the UAC to be certain that the feature is understood by the proxy. If the support of this option is desired but not required, it is listed in a Supported header field instead. An example is:

```
Proxy-Require: timer
```

6.2.15 P-OSP-Auth-Token

The P-OSP-Auth-Token header field [15] is used to transport an Open Settlements Protocol (OSP) token [16] with a SIP INVITE request. A gateway or proxy server receiving a token can verify the token and use this information about accepting the INVITE or rejecting the call. This approach is suitable for a clearinghouse model of VoIP carrier interconnection.

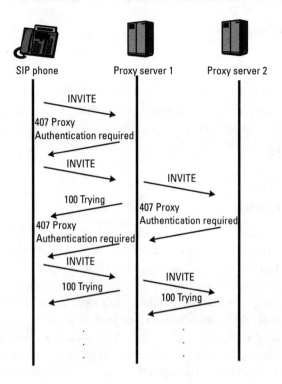

Figure 6.3 Multiproxy authentication example.

Table 6.15
Example of Proxy-Authorization Header Field

Header Field	Meaning
Proxy-Authorization: Digest username="Customer1", realm="company.com", nonce="9c8e88df84f1cec4341ae6e5a359", opaque="", uri="sip:user@company.com", response="e56131d19580cd833064787ecc"	This digest authorization header field contains the credentials of Customer1; the nonce was supplied by the SIP server located at the URI specified; the response contains the encrypted username and password; no opaque string is present

An example is:

```
P-OSP-Auth-Token: 3b8a40c10b4930ff19a85766c15182a34048d9398b834d6
  ;realm="carrier.com"
```

6.2.16 P-Asserted-Identity

The P-Asserted-Identity header field [17] is used between trusted intermediaries (proxies) to assert the identity of a user agent that has been authenticated using some means such as those described in Section 3.8. A UA receiving a request from a proxy that it trusts will typically render the value in a P-Asserted-Identity header field to the user as a "Verified Caller ID" as opposed to a From header value which is unverified. A proxy receiving a P-Asserted-Identity from another proxy that it does not trust will remove the header field.

An example is:

```
P-Asserted-Identity: <sip:user@example.com>
```

6.2.17 P-Preferred-Identity

The P-Preferred-Identity header field [17] is used by a user agent to tell a trusted intermediary which identity it would prefer be asserted on its behalf when more than one identities are associated with that user agent.

An example is:

```
P-Preferred-Identity: <sip:alternate@example.com>
```

6.2.18 Max-Forwards

The Max-Forwards header field is used to indicate the maximum number of hops that a SIP request may take. The value of the header field is decremented by each proxy that forwards the request. A proxy receiving the header field with a value of zero discards the message and sends a 483 Too Many Hops response back to the originator.

Max-Forwards is a mandatory header field in requests generated by a RFC 3261 compliant UA. However, an RFC 2543 UA generally will not include the header field field. The suggested initial value is 70 hops.

An example is:

```
Max-Forwards: 10
```

6.2.19 Reason

The Reason header field [18] can be used in BYE and CANCEL messages to indicate the reason why the session or call attempt is being terminated. It can carry a SIP response code or a Q.850 cause value (from an ISUP REL message, for example).

For example, a forking proxy could include the following header field in a CANCEL sent to a leg after one leg has answered the call:

```
Reason: SIP ;cause=200 ;text="Call completed elsewhere"
```

6.2.20 Refer-To

The `Refer-To` header field [19] is a required header field in a REFER request, which contains the URI or URL resource that is being referenced. It may contain any type of URI from a `sip` or `sips` to a `tel` URI to a `http` or `mailto` URI. For a `sip` or `sips` URI, the URI may contain a method or escaped header fields. For example, the following `Refer-To` header field:

```
Refer-To: <sip:UserC@client.anywhere.com?Replaces=
  sdjfdjfskdf@there.com%3Bto-tag%3D5f35a3%3Bfrom-tag%3D8675309>
```

contains an escaped `Replaces` header field. The resulting INVITE message generated by this `Refer-To` header field would have a Request-URI of `sip:UserC@client.anywhere.com` and a

```
Replaces: sdjfdjfskdf@there.com;to-tag=5f35a3;from-tag=8675309
```

header field. Note that the characters ; and = are replaced by their hex equivalents %3B and %3D. In the next example, the header field contains a method:

```
Refer-To: <sip:UserC@client.anywhere.com?method=SUBSCRIBE>
```

would cause a SUBSCRIBE request to be sent instead of an INVITE, which is the default method if none is present. An example of the `Refer-To` header field in compact form with an HTTP URL is:

```
r: <http://www.artech-house.com>
```

6.2.21 Referred-By

The `Referred-By` header field [20] is an optional header field in a REFER request and a request triggered by a REFER. It provides the recipient of a triggered request information that the request was generated as a result of a REFER and the originator of the REFER. This information can be presented to the user or have policy applied in deciding how the UA should handle the request. An unsigned `Referred-By` has the form:

```
Referred-By: <sip:user@host.com>
```

However, as this header field could be modified or fabricated, a more secure usage involves the addition of a `Referred-By` security token. The token is carried as a message body whose content id (`cid`) is indicated in the `Referred-By` header field. The token is an S/MIME signature over a `message/sipfrag`, which contains, at a minimum, the From, Date,

`Call-ID`, `Refer-To`, and `Referred-By` header fields from the REFER request. An example part of a REFER request is shown here:

```
Referred-By: sip:referror@referror.example
  ;cid=%3C39438823.FWQF33093@referror.example%3E
Content-Type: multipart/mixed; boundary=unique-boundary-1
Content-Length: (appropriate value)

—unique-boundary-1

Content-Type: multipart/signed;
 protocol="application/pkcs7-signature";
 micalg=sha1; boundary=divider42
Content-ID: <39438823.FWQF33093@referror.example>
Content-Length: (appropriate value)

—divider42
Content-Type: message/sipfrag
Content-Disposition: auth-id; handling=optional

From: sip:referror@referror.example
Date: Thu, 14 Feb 2003 16:23:03 GMT
Call-ID: 2203900ef0299349d9209f023a
Refer-To: sip:refertarget@target.example
Referred-By: sip:referror@referror.example
  ;cid=%3C39438823.FWQF33093@referror.example%3E

—divider42
Content-Type: application/pkcs7-signature; name=smime.p7s
Content-Transfer-Encoding: base64
Content-Disposition: attachment; filename=smime.p7s;
  handling=required

(appropriate signature goes here)

—divider42—

—unique-boundary-1—
```

An unsigned `Referred-By` header field may be rejected requesting that the `Referred-By` security token be included using the `429 Provide Referror Identity` response code (see Section 5.4.22). An example using the compact form is:

```
b: <sips:friend@neighbor.org>
```

6.2.22 Reply-To

The `Reply-To` header field is used to indicate a `sip` or `sips` URI, which should be used in replying to this request. Normally, this URI is present in the `From` header field (the `Contact` is not used as it is only assumed valid for the duration of the dialog). However, in some cases, the `From` cannot be populated with this information, so the URI in this header field should be used instead of the From URI.

An example is:

```
Reply-To: <sip:l.tolstoy@stpetersburg.ru>
```

6.2.23 Replaces

The Replaces header field [11] is used in SIP call control applications. A user agent in an established dialog receiving another INVITE with a Replaces header field that matches the existing dialog must accept the INVITE, terminate the existing dialog with a BYE, and transfer all resources and state from the existing dialog to the newly established dialog.

If the Replaces header field matches no dialog, the INVITE must be rejected with a 481 Dialog Does Not Exist response.

In addition, Replaces has one application in pending dialogs. A UAC that has sent an INVITE but has not yet received a final response may receive an INVITE containing a Replaces header field that matches the pending INVITE. The UAC must terminate the pending dialog with a CANCEL (and be prepared to send a ACK and BYE if a 200 OK eventually arrives) and accept the new INVITE.

For an INVITE containing both a Require: replaces and Replaces header field, this results in the return of one of the following set of responses:

- 200 (if a match is found);
- 481 (if no match is found);
- 420 (if Replaces is not supported).

Figure 6.4 shows a call flow using Replaces to implement a feature called "call pickup." Figure 4.7 shows the use of Replaces in an "attended transfer example."

This example Replaces header field:

```
Replaces: 3232904875945@server.org;to-tag=34314;from-tag=2343
```

would match the dialog identified by:

```
To: <sip:moe@example.org>;tag=34314
From: <sip:larry@server.org>;tag=2343
Call-ID: 3232904875945@server.org
```

6.2.24 Reject-Contact

The Reject-Contact [4] header field specifies the URIs to which the request may not be proxied. Some additional parameters are also defined for

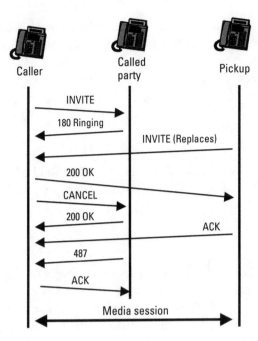

Figure 6.4 Call pickup call flow using `Replaces`.

`Contact` header fields such as `media,` `duplex`, and `language` when used in this header field. This header field, along with `Accept-Contact` and `Request-Disposition` are part of the SIP caller preferences extensions. The compact form is `j`. Examples include:

```
Reject-Contact: sip:admin@boss.com
j: *;media=video
```

6.2.25 Request-Disposition

The `Request-Disposition` [4] header field can be used to request servers to either proxy or redirect, or initiate serial or parallel (forking) searches. An example is:

```
Request-Disposition: redirect
```

6.2.26 Require

The `Require` header field is used to list features and extensions that a UAC requires a UAS to support in order to process the request. A `420 Bad Extension` response is returned by the UAS listing any unsupported features in an Unsupported header field. If support or use of a feature is desirable but not required, the Supported header field is used instead. See Table 6.8 for a list of feature tags.

An example is:

```
Require: rel100
```

6.2.27 Response-Key

The Response-Key header field was defined in RFC 2543 but was deprecated in RFC 3261 along with all PGP-based encryption in favor of S/MIME encryption.

6.2.28 Route

The Route header field is used to provide routing information for requests. RFC 3261 introduces two types of routing: strict and loose routing, which have similar meaning as the IP routing modes of the same name. In strict routing, a proxy must use the first URI in the Route header field to rewrite the Request-URI, which is then forwarded. In loose routing, a proxy does not re-write the Request-URI, but either forwards the request to the first URI in the Route header field or it may forward the request to another loose routing element. In loose routing, the request must route through every server in the Route list (but may also route through other servers) before it may be routed based on the Request-URI. In strict routing, the request must only route through the set of servers in the Route header field with the Request-URI being rewritten at each hop. A proxy or UAC can tell if the next element in the route set supports loose routing by the presence of a lr parameter. An example is:

```
Route: <sip:proxy@example.com;lr>
```

Chapter 10 contains an example of the use of the Record-Route and Route header fields. Examples of Route header fields constructed from the example Record-Route header fields in Section 6.2.12 are:

```
Route: <sip:firewall33.corporation.com;lr>,
  <sip:proxy1.carrier.com;lr>

Route: <sip:139.23.1.44 ;lr>
```

6.2.29 RAck

The RAck header field [10] is used within a response to a PRACK request to reliably acknowledge a provisional response that contained a RSeq header field. The RAck header field echoes the CSeq and the RSeq from the provisional response. The reliable sequence number is incremented for each response sent reliably. A call flow is shown in Figure 4.11. An example is:

```
RAck: 8342523 1 INVITE
```

6.2.30 Session-Expires

The Session-Expires header field [21] is used to specify the expiration time of the session. To extend the session, either UA can send a re-INVITE or UPDATE with a new Session-Expires header field. At the expiration of the interval in the Session-Expires, either UA may send a BYE and call-stateful proxies may destroy any state information. A proxy may shorten the expiration time by reducing the interval in the header field as it proxies the request. A UAS confirms the session timer by including the Session-Expires header field in the response to the request. A UAS may also shorten the interval by reducing the interval. An example is:

```
Session-Expires: 3600
```

6.2.31 Subscription-State

The Subscription-State header field [3] is a required header field in a NOTIFY request. It indicates the current state of the subscription. Values defined include active, pending, or terminated. Additional parameters include expires, reason, and retry-after. Values defined for the reason parameter include deactivated, giveup, probation, noresource, rejected, and timeout.

An example is:

```
Subscription-State: terminated ;reason=rejected
```

6.3 Response Header Fields

These header fields are present only in responses.

6.3.1 Authenticaton-Info

The Authentication-Info header field can be inserted in responses when performing mutual authentication using HTTP Digest. In normal HTTP Digest as described in Section 3.8, the server challenges the client to provide a shared secret, which the client then provides in a repeat of the request containing an Authorization or WWW-Authenticate header field. To do mutual authentication, the server would then provide an Authentication-Info header field containing either a next nonce or a response in an `repath` parameter. The response auth digest is calculated by the server on the SIP response using the same algorithm as the successful request authentication and using the same shared secret (client's username and password). In this way, the server proves that it also knows the client's secret, providing mutual authentication. The credentials are carried in the `rspauth` parameter in the header field.

An example is:

```
Authentication-Info: rspauth="9105jr981i459jgfp"
```

6.3.2   Error-Info

The `Error-Info` header field is used in failure response to convey more information about an error. A UAC receiving the header field in a failure response may fetch and render the URI to the user. The header field can be used to give the client the choice of how the error can be presented to the user. For example, a client with a graphical interface will likely display the reason phrase on the response, which should provide very specific information about the failure. However, an audio-only UA does not have this capability (although a text-to-speech synthesizer could be used to provide this capability). Instead, an audio-only UA could fetch the URI and play the resulting audio stream to the user.

If the URI is a `sip` or `sips` URI, the UA may treat the `Error-Info` as a `Contact` in a redirection response, which would result in a SIP session established to play the recording.

An example is:

```
Error-Info: <sip:recording5@announcementsrus.com>
```

6.3.3   Min-Expires

The `Min-Expires` header field is used in a `423 Interval Too Brief` response (Section 5.4.20) from a registrar rejecting a `REGISTER` request in which one or more `Contacts` have an expiration time that is too short. The header field contains an integer number of seconds that represents the minimum expiration interval that the registrar will accept. A client receiving this header field can update the expiration intervals of the registration request accordingly and resend the `REGISTER` request.

An example is:

```
Min-Expires: 1200
```

6.3.4   Min-SE

The `Min-SE` header field [21] is a required header field in a `422 Session Timer Interval Too Small` response (Section 5.4.19). The response may also be present in an INVITE or UPDATE containing a Session-Expires header field. It contains an integer number of seconds.

An example is:

```
Min-SE: 480
```

6.3.5 Proxy-Authenticate

The Proxy-Authenticate header field is used in a 407 Proxy Authentication Required authentication challenge by a proxy server to a UAC. It contains the nature of the challenge so that the UAC may formulate credentials in a Proxy-Authorization header field in a subsequent request. Examples are shown in Table 6.16.

6.3.6 Server

The Server header field is used to convey information about the UAS generating the response. The use and contents of the header field are similar to the User-Agent header field in Section 6.1.16. An example is:

```
Server: Dotcom Announcement Server B3
```

6.3.7 Unsupported

The Unsupported header field is used to indicate features that are not supported by the server. The header field is used in a 420 Bad Extension response to a request containing an unsupported feature listed in a Require header field. Because multiple features may have been listed in the Require header field, the Unsupported header field indicates all the unsupported features—the rest can be assumed by the UAC to be supported. See Table 6.8 for a list of feature tags.

An example is:

```
Unsupported: rel100
```

Table 6.16
Example of Proxy-Authenticate Header Field

Header Field	Meaning
Proxy-Authenticate: Digest realm="example.com", nonce="9c8e88df84f1cec4341ae6e5a359", opaque="", stale=FALSE, algorithm=MD5	HTTP digest challenge header field

6.3.8   Warning

The Warning header field is used in a response to provide more specific information than the response code alone can convey. The header field contains a three-digit warning code, a warning agent that indicates what server inserted the header field, and warning text enclosed in quotes used for display purposes. Warning codes in the 1xx and 2xx range are specific to HTTP [2]. The SIP standard defines 12 new warning codes in the 3xx class. The breakdown of the class is shown in Table 6.17. The complete set of defined warning codes is listed in Table 6.18.

Examples are:

```
Warning: 302 proxy "Incompatible transport protocol"
Warning: 305 room132.hotel.com:5060 "Incompatible media type"
```

6.3.9   WWW-Authenticate

The WWW-Authenticate header field is used in a 401 Unauthorized authentication challenge by a user agent or registrar server to a UAC. It contains the nature of the challenge so that the UAC may formulate credentials in a Proxy-Authorization header field in a subsequent request. SIP supports HTTP digest authentication mechanisms. Examples are shown in Table 6.19.

6.3.10   RSeq

The RSeq header field [10] is used in provisional (1xx class) responses to INVITEs to request reliable transport. The header field may only be used if the INVITE request contained the Supported: rel100 header field. If present in a provisional response, the UAC should acknowledge receipt of the

Table 6.17
SIP Warning Codes

Warning Code Range	Error Type
30x, 31x, 32x	SDP keywords
33x	Network services
34x, 35x, 36x	Reserved for future use
37x	QoS parameters
38x	Reserved
39x	Miscellaneous

Table 6.18
SIP Warning Code List

Warning Code	Description
300	Incompatible network protocol
301	Incompatible network address formats
302	Incompatible transport protocol
303	Incompatible bandwidth units
304	Media type not available
305	Incompatible media format
306	Attribute not understood
307	Session description parameter not understood
330	Multicast not available
331	Unicast not available
370	Insufficient bandwidth
399	Miscellaneous warning

Table 6.19
Example of WWW-Authenticate Header Field

Header Field	Meaning
WWW-Authenticate: Digest realm="example.com", nonce="9c8e88df84f1cec4341ae6e5a359", opaque="", stale=FALSE, algorithm=MD5	HTTP digest challenge

response with a PRACK method, as described in Section 4.1.12. The RSeq header field contains a reliable sequence number that is an integer randomly initialized by the UAS. Each subsequent provisional response sent reliably for this dialog will have a monotonically increasing RSeq number. The UAS matches the reliable sequence number and CSeq from the RAck in a PRACK request to a sent response to confirm receipt and stop all retransmissions of the response.

An example is:

```
RSeq: 2345263
```

6.4   Message Body Header Fields

These header fields contain information about the message body or entity.

6.4.1   Allow

The `Allow` header field is used to indicate the methods supported by the user agent or proxy server sending the response. The header field must be present in a `405 Method Not Allowed` response and should be included in a positive response to an `OPTIONS` request. An example is:

```
Allow: INVITE, ACK, BYE, INFO, OPTIONS, CANCEL
```

6.4.2   Content-Encoding

The `Content-Encoding` header field is used to indicate that the listed encoding scheme has been applied to the message body. This allows the UAS to determine the decoding scheme necessary to interpret the message body. Multiple listings in this header field indicate that multiple encodings have been used in the sequence in which they are listed. Only encoding schemes listed in an `Allow-Encoding` header field may be used. The compact form is `e`. Examples include:

```
Content-Encoding: text/plain
e: gzip
```

6.4.3   Content-Disposition

The `Content-Disposition` header field is used to describe the function of a message body. Defined values include `session, icon, alert`, and `render`. The value session indicates that the message body contains information to describe a media session. The value render indicates that the message body should be displayed or otherwise rendered for the user. If a message body is present in a request or a 2xx response without a `Content-Disposition`, the function is assumed to be session. For all other response classes with message bodies, the default function is render. An example is:

```
Content-Function: session
```

6.4.4   Content-Language

The `Content-Language` header field [2] is used to indicate the language of a message body. It contains a language tag, which identifies the language.

```
Content-Language: en
```

6.4.5   Content-Length

The `Content-Length` is used to indicate the number of octets in the message body. A `Content-Length: 0` indicates no message body. As described in Section 2.4.2, this header field is used to separate multiple messages sent within a TCP stream. If not present in a UDP message, the message body is assumed to continue to the end of the datagram. If not present in a TCP message, the message body is assumed to continue until the connection is closed. The `Content-Length` octet count does not include the CRLF that separates the message header fields from the message body. It does, however, include the CRLF at the end of each line of the message body. An example octet calculation is in Chapter 2. The `Content-Length` header field is not a required header field to allow dynamically generated message bodies, where the `Content-Length` may not be known a priori. The compact form is `l`. Examples include:

```
Content-Length: 0
l: 287
```

6.4.6   Content-Type

The `Content-Type` header field is used to specify the Internet media type [13] in the message body. Media types have the familiar form `type/sub-type`. If this header field is not present, `application/sdp` is assumed. If an `Accept` header field was present in the request, the response `Content-Type` must contain a listed type, or a `415 Unsupported Media Type` response must be returned. The compact form is `c`. Specific MIME types that are commonly use are listed in Table 6.20.

Content indirection [28] can be used to provide a URI in place of actual MIME message body. An example is:

Table 6.20
Common Content-Types Present in SIP Requests and Responses

Content-Type	Use
`application/sdp`	SDP in INVITE, ACK, or UPDATE requests [22]
`message/sipfrag`	SIP fragment in NOTIFY in refer subscription [23]
`application/xml+dialog`	XML dialog [24]
`application/xml+conf`	XML conference info [25]
`application/cpim`	CPIM [26]
`text/plain`	Plain text
`text/html`	HTML text
`application/isup`	Encapsulated ISUP in INVITE, BYE, or INFO [27]

```
Content-Type: message/external-body; access-type="URL";
 URL="http://www.example.com/"
```

The compact form is c. Examples are:

```
Content-Type: application/sdp
c: text/html
```

6.4.7   Expires

The Expires header field is used to indicate the time interval in which the request or message contents are valid. When present in an INVITE request, the header field sets a time limit on the completion of the INVITE request. That is, the UAC must receive a final response (non-1xx) within the time period or the INVITE request is automatically canceled with a 408 Request Timeout response. Once the session is established, the value from the Expires header field in the original INVITE has no effect—the Session-Expires header field (Section 6.2.19) must be used for this purpose. When present in a REGISTER request, the header field sets the time limit on the URIs in Contact header fields that do not contain an expires parameter. Table 4.3 shows examples of the Expires header field in registration requests. The header field is not defined for any other method types. The header field may contain a SIP date or a number of seconds. Examples include:

```
Expires: 60
Expires: Fri, 15 Apr 2000 00:00:00 GMT
```

6.4.8   MIME-Version

The MIME-Version header field is used to indicate the version of MIME protocol used to construct the message body. SIP, like HTTP, is not considered MIME-compliant because parsing and semantics are defined by the SIP standard, not the MIME Specification [29]. Version 1.0 is the default value. An example is:

```
MIME-Version: 1.0
```

References

[1]   Rosenberg, J., et al., "SIP: Session Initiation Protocol," RFC 3261, 2002.

[2]   Fielding, R., et al., "Hypertext Transfer Protocol — HTTP/1.1," RFC 2616, June 1999.

[3]   Roach, A., "Session Initiation Protocol (SIP)-Specific Event Notification," RFC 3265, 2002.

[4] Rosenberg, J., H. Schulzrinne, and P. Kyzivat, "Caller Preferences and Callee Capabilities for the Session Initiation Protocol (SIP)," IETF Internet-Draft, Work in Progress, March 2003.

[5] Johnston, A., and O. Levin, "Session Initiation Protocol (SIP) Call Control – Conferencing for User Agents," IETF Internet-Draft, Work in Progress, April 2003.

[6] Rosenberg, J., and H. Schulzrinne, "A Session Initiation Protocol (SIP) Event Package for Conference State," IETF Internet-Draft, Work in Progress, June 2002.

[7] Braden, R., "Requirements for Internet Hosts: Application and Support," RFC 1123, 1989.

[8] Mahy, R., and D. Petrie, "The Session Inititation Protocol (SIP) 'Join' Header," IETF Internet-Draft, Work in Progress, March 2003.

[9] Willis, D., and B. Hoeneisen, "Session Initiation Protocol (SIP) Extension Header Field for Registering Non-Adjacent Contacts," RFC 3327, 2002.

[10] Rosenberg, J., and H. Schulzrinne, "Reliability of Provisional Responses in Session Initiation Protocol (SIP)," RFC 3262, 2002.

[11] Mahy, R., B. Biggs, and R. Dean, "The Session Initiation Protocol (SIP) 'Replaces' Header," IETF Internet-Draft, Work in Progress, March 2003.

[12] Donovan, S., and J. Rosenberg, "Session Timers in the Session Initiation Protocol (SIP)," IETF Internet-Draft, Work in Progress, November 2002.

[13] Postel, J., "Media Type Registration Procedure," RFC 1590, 1994.

[14] Peterson, J., "A Privacy Mechanism for the Session Initiation Protocol (SIP)," RFC 3323, 2002.

[15] Johnston, A., et al., "Session Initiation Protocol Private Extension for an OSP Authorization Token," IETF Internet-Draft, Work in Progress, February 2003.

[16] European Telecommunications Standards Institute, "Telecommunications and Internet Protocol Harmonization Over Networks (TIPHON); Open Settlement Protocol (OSP) for Inter-domain Pricing, Authorization, and Usage Exchange," Technical Specification 101 321, Version 2.1.0.

[17] Jennings, C., J. Peterson, and M. Watson, "Private Extensions to the Session Initiation Protocol (SIP) for Asserted Identity within Trusted Networks," RFC 3325, 2002.

[18] Schulzrinne, H., D. Oran, and G. Camarillo, "The Reason Header Field for the Session Initiation Protocol (SIP)," RFC 3326, 2002.

[19] Sparks, R., "The Session Initiation Protocol (SIP) Refer Method," RFC 3515, 2003.

[20] Sparks, R., "The SIP Referred-By Mechanism," IETF Internet-Draft, Work in Progress, February 2003.

[21] Donovan, S., and Rosenberg, J., "Session Timers in the Session Initiation Protocol (SIP)," IETF Internet-Draft, Work in Progress, November 2002.

[22] Handley, M., and V. Jacobson, "SDP: Session Description Protocol," RFC 2327, 1998.

[23] Sparks, R., "Internet Media Type message/sipfrag," RFC 3420, 2003.

[24] Rosenberg, J., and H. Schulzrinne, "An INVITE Initiated Dialog Event Package for the Session Initiation Protocol (SIP)," IETF Internet-Draft, Work in Progress, March 2003.

[25] Rosenberg, J., and H. Schulzrinne, "A Session Initiation Protocol (SIP) Event Package for Conference State," IETF Internet-Draft, Work in Progress, June 2002.

[26]   Sugano, H., et al., "Common Presence and Instant Messaging (CPIM) Presence Information Data Format," IETF Internet-Draft, Work in Progress, 2002

[27]   Zimmerer, E., et al., "MIME Media Types for ISUP and QSIG Objects," RFC 3204, December 2001.

[28]   Olson, S., "A Mechanism for Content Indirection in Session Initiation Protocol (SIP) Messages," IETF Internet-Draft, Work in Progress, February 2003.

[29]   Freed, M., and N. Borenstein, "Multipurpose Internet Mail Extensions (MIME). Part One: Format of Internet Message Bodies," RFC 2045, 1996.

7

Related Protocols

The Session Initiation Protocol (SIP) is one part of the protocol suite that makes up the Internet Multimedia Conferencing architecture, as shown in Figure 1.1. In this chapter, other related Internet protocols mentioned or referenced in other sections are introduced, along with details on the use of the protocol with SIP. This is by no means a complete discussion of multimedia communication protocols over the Internet. First, SDP, the media description language, will be discussed. Then the RTP and RTCP media transport protocols will be discussed. The application of RTP/AVP profiles that link SDP and RTP will then be then covered. The chapter concludes with a brief discussion of signaling protocols in the PSTN. The H.323 protocol will be discussed and compared to SIP in the next chapter. The chapter concludes with a discussion of SIP-T and UPnP protocols.

7.1 SDP—Session Description Protocol

The Session Description Protocol, defined by RFC 2327 [1], was developed by the IETF MMUSIC working group. It is more of a description syntax than a protocol in that it does not provide a full-range media negotiation capability. The original purpose of SDP was to describe multicast sessions set up over the Internet's multicast backbone, the MBONE. The first application of SDP was by the experimental Session Announcement Protocol (SAP) [2] used to post and retrieve announcements of MBONE sessions. SAP messages carry a SDP message body, and was the template for SIP's use of SDP. Even though it was designed for multicast, SDP has been applied to the more general problem of describing general multimedia sessions established using SIP.

As seen in the examples of Chapter 3, SDP contains the following information about the media session:

- IP Address (IPv4 address or host name);
- Port number (used by UDP or TCP for transport);
- Media type (audio, video, interactive whiteboard, and so forth);
- Media encoding scheme (PCM A-Law, MPEG II video, and so forth).

In addition, SDP contains information about the following:

- Subject of the session;
- Start and stop times;
- Contact information about the session.

Like SIP, SDP uses text coding. An SDP message is composed of a series of lines, called fields, whose names are abbreviated by a single lower-case letter, and are in a required order to simplify parsing. The set of mandatory SDP fields is shown in Table 2.1. The complete set is shown in Table 7.1.

SDP was not designed to be easily extensible, and parsing rules are strict. The only way to extend or add new capabilities to SDP is to define a new attribute type. However, unknown attribute types can be silently ignored. A SDP parser must not ignore an unknown field, a missing mandatory field, or an out-of-sequence line. An example SDP message containing many of the optional fields is shown here:

```
v=0
o=johnston 2890844526 2890844526 IN IP4 43.32.1.5
s=SIP Tutorial
i=This broadcast will cover this new IETF protocol
u=http://www.digitalari.com/sip
e=Alan Johnston alan@mci.com
p=+1-314-555-3333 (Daytime Only)
c=IN IP4 225.45.3.56/236
b=CT:144
t=2877631875 2879633673
m=audio 49172 RTP/AVP 0
a=rtpmap:0 PCMU/8000
m=video 23422 RTP/AVP 31
a=rtpmap:31 H261/90000
```

The general form of a SDP message is:

```
x=parameter1 parameter2 ... parameterN
```

Table 7.1
SDP Field List in Their Required Order

Field	Name	Mandatory/Optional
v=	Protocol version number	m
o=	Owner/creator and session identifier	m
s=	Session name	m
i=	Session information	o
u=	Uniform Resource Identifer	o
e=	Email address	o
p=	Phone number	o
c=	Connection information	m
b=	Bandwidth information	o
t=	Time session starts and stops	m
r=	Repeat times	o
z=	Time zone corrections	o
k=	Encryption key	o
a=	Attribute lines	o
m=	Media information	o
a=	Media attributes	o

The line begins with a single lower-case letter x. There are never any spaces between the letter and the =, and there is exactly one space between each parameter. Each field has a defined number of parameters. Each line ends with a CRLF. The individual fields will now be discussed in detail.

7.1.1 Protocol Version

The v= field contains the SDP version number. Because the current version of SDP is 0, a valid SDP message will always begin with v=0.

7.1.2 Origin

The o= field contains information about the originator of the session and session identifiers. This field is used to uniquely identify the session. The field contains:

```
o=username session-id version network-type address-type
  address
```

The `username` parameter contains the originator's login or host or - if none. The `session-id` parameter is a Network Time Protocol (NTP) [3] timestamp or a random number used to ensure uniqueness. The `version` is a numeric field that is increased for each change to the session, also recommended to be a NTP timestamp. The `network-type` is always `IN` for Internet. The `address-type` parameter is either `IP4` or `IP6` for IPv4 or IPv6 address either in dotted decimal form or a fully qualified host name.

7.1.3   Session Name and Information

The `s=` field contains a name for the session. It can contain any non-zero number of characters. The optional `i=` field contains information about the session. It can contain any number of characters.

7.1.4   URI

The optional `u=` field contains a uniform resource indicator (URI) with more information about the session.

7.1.5   E-Mail Address and Phone Number

The optional `e=` field contains an e-mail address of the host of the session. If a display name is used, the e-mail address is enclosed in <>. The optional `p=` field contains a phone number. The phone number should be given in globalized format, beginning with a +, then the country code, a space or -, then the local number. Either spaces or - are permitted as spacers in SDP. A comment may be present in ().

7.1.6   Connection Data

The `c=` field contains information about the media connection. The field contains:

```
c=network-type address-type connection-address
```

The network-type parameter is defined as `IN` for the Internet. The address type is defined as `IP4` for IPv4 addresses, `IP6` for IPv6 addresses. The connection-address is the IP address that will be sending the media packets, which could be either multicast or unicast. If multicast, the connection-address field contains:

```
connection-address=base-multicast-address/ttl/number-of-
   addresses
```

where `ttl` is the time-to-live value, and `number-of-addresses` indicates how many contiguous multicast addresses are included starting with the `base-multicast-address`.

7.1.7 Bandwidth

The optional `b=` field contains information about the bandwidth required. It is of the form:

```
b=modifier:bandwidth-value
```

The `modifier` is either `CT` for conference total or `AS` for application specific. `CT` is used for multicast session to specify the total bandwidth that can be used by all participants in the session. `AS` is used to specify the bandwidth of a single site. The `bandwidth-value` parameter is the specified number of kilobytes per second.

7.1.8 Time, Repeat Times, and Time Zones

The `t=` field contains the start time and stop time of the session.

```
t=start-time stop-time
```

The times are specified using NTP timestamps. For a scheduled session, a `stop-time` of zero indicates that the session goes on indefinitely. A `start-time` and `stop-time` of zero for a scheduled session indicates that it is permanent. The optional `r=` field contains information about the repeat times that can be specified in either in NTP or in days (d), hours (h), or minutes (m). The optional `z=` field contains information about the time zone offsets. This field is used if a reoccurring session spans a change from daylight-savings to standard time, or vice versa.

7.1.9 Encryption Keys

The optional `k=` field contains the encryption key to be used for the media session. The field contains:

```
k=method:encryption-key
```

The `method` parameter can be `clear`, `base64`, `uri`, or `prompt`. If the method is `prompt`, the key will not be carried in SDP; instead, the user will be prompted as they join the encrypted session. Otherwise, the key is sent in the `encryption-key` parameter.

7.1.10 Media Announcements

The optional m= field contains information about the type of media session. The field contains:

```
m=media port transport format-list
```

The media parameter is either `audio`, `video`, `application`, `data`, `telephone-event`, or `control`. The `port` parameter contains the port number. The `transport` parameter contains the transport protocol, which is either `RTP/AVP` or udp. (RTP/AVP stands for Real-time Transport Protocol [4] / audio video profiles [5], which is described in Section 7.3.) The `format-list` contains more information about the media. Usually, it contains media payload types defined in RTP audio video profiles. More than one media payload type can be listed, allowing multiple alternative codecs for the media session. For example, the following media field lists three codecs:

```
m=audio 49430 RTP/AVP 0 6 8 99
```

One of these three codecs can be used for the audio media session. If the intention is to establish three audio channels, three separate media fields would be used. For non-RTP media, Internet media types should be listed in the `format-list`. For example,

```
m=application 52341 udp wb
```

could be used to specify the `application`/wb media type.

7.1.11 Attributes

The optional a= field contains attributes of the preceding media session. This field can be used to extend SDP to provide more information about the media. If not fully understood by a SDP user, the attribute field can be ignored. There can be one or more attribute fields for each media payload type listed in the media field. For the RTP/AVP example in Section 7.1.10, the following three attribute fields could follow the media field:

```
a=rtpmap:0 PCMU/8000
a=rtpmap:6 DVI4/16000
a=rtpmap:8 PCMA/8000
a=rtpmap:99 iLBC
```

Other attributes are shown in Table 7.2. Full details of the use of these attributes are in the standard document [1]. The details of the iLBC (Internet Low Bit Rate) Codec are in [6].

7.1.12 Use of SDP in SIP

The use of SDP with SIP is given in the SDP Offer Answer RFC 3264 [7]. The default message body type in SIP is application/sdp. The calling party lists the media capabilities that they are willing to receive in SDP in either an INVITE or in an ACK. The called party lists their media capabilities in the 200 OK response to the INVITE. More generally, offers or answers may be in INVITEs, PRACKs, or UPDATEs or in reliably sent 18x or 200 responses to these methods.

Because SDP was developed with scheduled multicast sessions in mind, many of the fields have little or no meaning in the context of dynamic sessions established using SIP. In order to maintain compatibility with the SDP protocol, however, all required fields are included. A typical SIP use of SDP includes the version, origin, subject, time, connection, and one or more media and attribute fields as shown in Table 2.1. The origin, subject, and time fields are not used by SIP but are included for compatibility. In the SDP standard, the subject field is a required field and must contain at least one character, suggested to be s=- if there is no subject. The time field is usually set to t=0 0.

SIP uses the connection, media, and attribute fields to set up sessions between user agents. Because the type of media session and codec to be used are part of the connection negotiation, SIP can use SDP to specify multiple alternative media types and to selectively accept or decline those media types. When multiple media codecs are listed, the caller and called party's media fields must be aligned—that is, there must be the same number, and they must be listed in the same order. The offer answer specification, RFC 3264 [7], recommends that an attribute containing a=rtpmap: be used for each media field [7]. A media stream is declined by setting the port number to zero for the corresponding media field in the SDP response. In the following example, the caller Tesla wants to set up an audio and video call with two possible audio codecs and a video codec in the SDP carried in the initial INVITE:

```
v=0
o=Tesla 2890844526 2890844526 IN IP4 lab.high-voltage.org
s=-
c=IN IP4 100.101.102.103
t=0 0
m=audio 49170 RTP/AVP 0 8
a=rtpmap:0 PCMU/8000
a=rtpmap:8 PCMA/8000
```

Table 7.2
SDP Attribute Values

Attribute	Name
a=rtpmap:	RTP/AVP list
a=cat:	Category of the session
a=keywds:	Keywords of session
a=tool:	Name of tool used to create SDP
a=ptime:	Length of time in milliseconds for each packet
a=recvonly	Receive only mode
a=sendrecv	Send and receive mode
a=sendonly	Send only mode
a=orient:	Orientation for whiteboard sessions
a=type:	Type of conference
a=charset:	Character set used for subject and information fields
a=sdplang:	Language for the session description
a=lang:	Default language for the session
a=framerate:	Maximum video frame rate in frames per second
a=quality:	Suggests quality of encoding
a=fmtp:	Format transport
a=mid:	Media identification grouping
a=direction:	Direction for symmetric media
a=rtcp:	Explicit RTCP port (and address)
a=inactive	Inactive mode

```
m=video 49172 RTP/AVP 32
a=rtpmap:32 MPV/90000
```

The codecs are referenced by the RTP/AVP profile numbers 0, 8, and 32. The called party Marconi answers the call, chooses the second codec for the first media field and declines the second media field, only wanting a PCM A-Law audio session.

```
v=0
o=Marconi 2890844526 2890844526 IN IP4 tower.radio.org
s=-
c=IN IP4 200.201.202.203
```

```
t=0  0
m=audio 60000 RTP/AVP 8
a=rtpmap:8 PCMA/8000
m=video 0 RTP/AVP 32
```

If this audio-only call is not acceptable, then Tesla would send an ACK then a BYE to cancel the call. Otherwise, the audio session would be established and RTP packets exchanged. As this example illustrates, unless the number and order of media fields is maintained, the calling party would not know for certain which media sessions were being accepted and declined by the called party.

One party in a call can temporarily place the other on hold (i.e., suspending the media packet sending). This is done by sending an INVITE with identical SDP to that of the original INVITE but with a=sendonly attribute present. The call is made active again by sending another INVITE with the a=sendrecv attribute present. (Note that older RFC 2543 compliant UAs may initiate hold using c=0.0.0.0.) For further examples of SDP use with SIP, see the SDP Offer Answer Examples document [8].

7.2 RTP—Real-Time Transport Protocol

Real-Time Transport Protocol [4] was developed to enable the transport of real-time packets containing voice, video, or other information over IP. RTP is defined by IETF Proposed Standard RFC 3550 (which updates RFC 1889). RTP does not provide any quality of service over the IP network—RTP packets are handled the same as all other packets in an IP network. However, RTP allows for the detection of some of the impairments introduced by an IP network, such as:

- Packet loss;
- Variable transport delay;
- Out of sequence packet arrival;
- Asymmetric routing.

As shown in the protocol stack of Figure 1.1, RTP is an application layer protocol that uses UDP for transport over IP. RTP is not text encoded, but uses a bit-oriented header similar to UDP and IP. RTP version 0 is only used by the *vat* audio tool for MBONE broadcasts. Version 1 was a pre-RFC implementation and is not in use. The current RTP version 2 packet header is shown in Figure 7.1. RTP was designed to be very general; most of the headers are only loosely defined in the standard; the details are left to profile documents. The 12 octets are defined as:

V	P	X	CC	M	PT	Sequence Number	Timestamp	SSRCI

Figure 7.1 RTP packet header.

- Version (V): This 2-bit field is set to 2, the current version of RTP.
- Padding (P): If this bit is set, there are padding octets added to the end of the packet to make the packet a fixed length. This is most commonly used if the media stream is encrypted.
- Extension (X): If this bit is set, there is one additional extension following the header (giving a total header length of 14 octets). Extensions are defined by certain payload types.
- CSRC count (CC): This 4-bit field contains the number of content source identifiers (CSRC) are present following the header. This field is only used by mixers that take multiple RTP streams and output a single RTP stream.
- Marker (M): This single bit is used to indicate the start of a new frame in video, or the start of a talk-spurt in silence-suppressed speech.
- Payload Type (PT): This 7 bit field defines the codec in use. The value of this field matches the profile number listed in the SDP.
- Sequence Number: This 16-bit field is incremented for each RTP packet sent and is used to detect missing/out of sequence packets.
- Timestamp: This 32-bit field indicates in relative terms the time when the payload was sampled. This field allows the receiver to remove jitter and to play back the packets at the right interval assuming sufficient buffering.
- Synchronization Source Identifier (SSRCI): This 32-bit field identifies the sender of the RTP packet. At the start of a session, each participant chooses a SSRC number randomly. Should two participants choose the same number, they each choose again until each party is unique.
- CSRC Contributing Source Identifier: There can be none or up to 15 instances of this 32-bit field in the header. The number is set by the CSRC Count (CC) header field. This field is only present if the RTP packet is being sent by a mixer, which has received RTP packets from a number of sources and sends out combined packets. A non-multicast conference bridge would utilize this header.

RTP allows detection of a lost packet by a gap in the Sequence Number. Packets received out of sequence can be detected by out-of-sequence

Sequence Numbers. Note that RTP allows detection of these transport-related problems but leaves it up to the codec to deal with the problem. For example, a video codec may compensate for the loss of a packet by repeating the last video frame, while an audio codec may play background noise for the interval. Variable delay or jitter can be detected by the Timestamp field. A continuous bit rate codec such as PCM will have a linearly increasing Timestamp. A variable bit rate codec, however, which sends packets at irregular intervals, will have an irregularly increasing Timestamp, which can be used to play back the packets at the correct interval.

The RTP Control Protocol (RTCP) is a related protocol also defined in RFC 3550 that allows participants in an RTP session to send each other quality reports and statistics, and exchange some basic identity information. The four types of RTCP packets are shown in Table 7.3. RTCP has been designed to scale to very large conferences. Because RTCP traffic is all overhead, the bandwidth allocated to these messages remains fixed regardless of the number of participants. That is, the more participants on a conference, the less frequently RTCP packets are sent. For example, in a basic two-participant audio RTP session, the RTP/AVP profile states that RTCP packets are to be sent about every 5 seconds; for four participants, RTCP packets can be sent every 10 seconds. Sender reports (SR) or receiver reports (RR) packets are sent the most frequently, with the other packet types being sent less frequently. The use of reports allows feedback on the quality of the connection including information such as:

- Number of packets sent and received;
- Number of packets lost;
- Packet jitter.

Table 7.3
RTCP Packet Types

Packet Type	Name	Description
SR	Sender report	Sent by a participant that both sends and receives RTP packets
RR	Receiver report	Sent by a participant that only receives RTP packets
SDES	Source description	Contain information about the participant in the session including e-mail address, phone number, and host
BYE	Bye	Sent to terminate the RTP session
APP	Application specific	Defined by a particular profile
XR	Extended report	Extended report and summaries

In a multimedia session established with SIP, the information needed to select codecs and send the RTP packets to the right location is carried in the SDP message body. Under some scenarios, it can be desirable to change codecs during an RTP session. An example of this relates to the transport of dual tone multiple frequency (DTMF) digits. A low bit rate codec that is optimized for transmitting vocal sounds will not transport the superimposed sine waves of a DTMF signal without introducing significant noise, which may cause the DTMF digit receiver to fail to detect the digit. As a result, it is useful to switch to another codec when the sender detects a DTMF tone. Because a RTP packet contains the payload type, it is possible to change codecs "on the fly" without any signaling information being exchanged between the user agents. On the other hand, switching codecs in general should probably not be done without a SIP signaling exchange (re-INVITE) beacuse the call could fail if one side switches to a codec that the other does not support. The SIP re-INVITE message exchange allows this change in media session parameters to fail without causing the established session to fail.

The use of random numbers for CSRC provides a minimal amount of security against "media spamming" where a literally *uninvited* third party tries to break into a media session by sending RTP packets during an established call. Unless the third party can guess the CSRC of the intended sender, the receiver will detect a change in CSRC number and either ignore the packets or inform the user that something is going on. This behavior for RTP clients, however, is not universally accepted, because in some scenarios (wireless hand-off, announcement server, call center, and so forth.) it might be desirable to send media from multiple sources during the progress of a call.

RTP supports encryption of the media. In addition, RTP can use IPSec [9] for authentication and encryption.

7.3   RTP Audio Video Profiles

The use of profiles enables RTP to be an extremely general media transport protocol. The current audio video profiles defined by RFC 3551 are listed in Table 7.4. The profile document makes the following specifications for RTP:

- UDP is used for underlying transport;
- RTP port numbers are always even, the corresponding RTCP port number is the next highest port, always an odd number;
- No header extensions are used.

For each of the profiles listed in Table 7.4, the profile document lists details of the codec, or a reference for the details is provided. Payloads in the

Table 7.4
RTP/AVP Audio and Video Payload Types

Payload	Codec	Clock	Description
0	PCMU	8000	ITU G.711 PCM μ-Law Audio 64 Kbps
1	1016	8000	CELP Audio 4.8 Kbps
2	G721	8000	ITU G721 ADPCM Audio 32 Kbps
3	GSM	8000	European GSM Audio 13 Kbps
5	DVI4	8000	DVI ADPCM Audio 32 Kbps
6	DVI4	16000	DVI ADPCM 64 Kbps
7	LPC	8000	Experimental LPC Audio
8	PCMA	8000	ITU G.711 PCM A-Law Audio 64 Kbps
9	G722	8000	ITU G.722 Audio
10	L16	44100	Linear 16-bit Audio 705.6 Kbps
11	L16	44100	Linear 16-bit Stereo Audio 1411.2 Kbps
14	MPA	90000	MPEG-I or MPEG-II Audio Only
15	G728	8000	ITU G.728 Audio 16 Kbps
25	CELB	90000	CelB Video
26	JBEG	90000	JBEG Video
28	NV	90000	nv Video
31	H261	90000	ITU H.261 Video
32	MPV	90000	MPEG-I and MPEG-II Video
33	MP2T	90000	MPEG-II transport stream Video
dynamic	iLBC	—	Internet low bit rate 15 Kbps [6]
dynamic	AMR	—	Adaptive Multirate Codec [12]

range 96–127 can be defined dynamically during a session. The minimum payload support is defined as 0 (PCMU) and 5 (DVI4). The document recommends dynamically assigned port numbers, although 5004 and 5005 have been registered for use of the profile and can be used instead. The standard also describes the process of registering new payload types with IANA. There are other references for a tutorial description of many of these audio codecs [10] and video codecs [11].

The information in the first three columns of Table 7.4 is also contained in the SDP `a=rtpmap:` field, which is why the attribute is optional.

7.4 PSTN Protocols

Three types of PSTN signaling protocols are mentioned in this text: Circuit Associated Signaling (CAS), ISDN (Integrated Services Digital Network), and ISUP (ISDN User Part). They will be briefly introduced and explained. How these protocols work in the PSTN today are covered in other references [10].

7.4.1 Circuit Associated Signaling

This type of signaling is the oldest currently used in the PSTN today. The signaling information uses the same audio circuit as the voice path, with digits and characters represented by audio tones. These are the tones that used to be barely discernible on long-distance calls before ring tone is heard. The tones are called multifrequency (MF) tones. They are similar to the tones used to signal between a telephone and a central office switch, which are DTMF tones. Long post dial delay is introduced because of the time taken to out-pulse long strings of digits. Also, CAS is susceptible to fraud, as fraudulent tones can be generated by the caller to make free telephone calls. This type of signaling is common in trunk circuits between a central office and a corporation's private branch exchange (PBX) switch.

7.4.2 ISUP Signaling

ISDN User Part is the protocol used between telephone switches in the PSTN for call signaling. It is used over a dedicated packet-switched network that uses Signaling System #7 (SS7) for transport. This signaling method was developed to overcome some of the delay and security problems with CAS. There are examples of ISUP signaling in the call flow examples of Chapter 10. The adoption of this out-of-band signaling protocol was the first step taken by telecommunications carriers away from circuit-switched networks and towards packet-switched networks. The final step will be moving the bearer channels onto a packet-switched network.

7.4.3 ISDN Signaling

Integrated Services Digital Network (ISDN) signaling was developed for all-digital telephone connections to the PSTN. The most common types of interfaces are the basic rate interface (BRI) and the primary rate interface (PRI). A BRI can contain two 64-Kbps B-channels for either voice or data and a 16-Kbps D-channel for signaling. BRI can be used over conventional telephone lines but requires an ISDN telephone or terminal adapter. PRI uses a 1.544-Mbps link called a T-1 or a DS-1, which is divided up into 23 B-channels and one D-channel, with each channel being 64 Kbps. The H.323 protocol, described in

Chapter 9, reuses a subset of the ISDN Q.931 signaling protocol used over the D-channel.

7.5 SIP for Telephones

SIP for Telephones (SIP-T) is a framework for SIP interworking with the PSTN [13]. It includes two approaches: translation and encapsulation. Both approaches will be discussed.

Translation is the direct mapping between PSTN protocols and SIP. The mapping between common PSTN protocols such as ISUP [14], Q.SIG [15], and others have been defined. Examples of SIP interworking with PSTN protocols including ISDN and CAS are in the SIP PSTN Call Flows document [16]. In this approach, as much of the information that is common to each protocol are mapped between them, with the remaining values being set to configurable defaults. In this approach, a SIP call from a PSTN gateway is indistinguishable from a SIP call from a native device, and is handled such by the protocol. However, since not every single parameter in a PSTN signaling message has a counterpart (or has any meaning) in SIP, some information is lost if the call routes back to a PSTN termination point.

Encapsulation is another approach that is only useful for SIP/PSTN gateways. Using this approach, first, PSTN-to-SIP translation is done to construct the appropriate SIP message, then the PSTN protocol message is encapsulated and included with the SIP message as a message body. If the SIP message is received by another SIP/PSTN gateway, the resulting PSTN signaling message is constructed from both the SIP message *and* the encapsulated PSTN message that was received by the other gateway. This approach offers the possibility of transparency (i.e., no loss of PSTN information as a call is carried across a SIP network). However, this only works in a network in which only one variation of PSTN protocol is used. Unlike Internet protocols, PSTN protocols vary by region and are not compatible without a special type of PSTN switch, which converts one message format to another. There are many dozens of protocol variants used throughout the world.

Another disadvantage of encapsulation is that it the PSTN message bodies must be encrypted if they are transported over the public Internet, or used in a network with native SIP devices. This is because private information can be carried in PSTN messages because PSTN protocols assume a different trust model than an Internet protocol such as SIP. To prevent accidental disclosure of this information, the message bodies must be encrypted by the originating gateway and decrypted by the terminating gateway, which adds significant processing requirements and call setup delay.

Encapsulated PSTN messages are carried as MIME bodies, which have been standardized for both ISUP and QSIG [17].

7.6   Universal Plug and Play Protocol

The Universal Plug and Play Protocol (UPnP) [18] is a protocol developed to allow configuration and peer-to-peer networking of intelligent IP devices. UPnP works together with DHCP or Auto IP to configure and set up devices in small networks typical of home and small office networks [19]. The protocol has six basic functions: addressing, discovery, description, control, events, and presentation.

Some SIP clients use UPnP to gain configuration information and to talk to firewalls and NATs that are UPnP enabled to open firewalls and learn private/public IP address mappings. For example, if a router or wireless bridge that encompasses a firewall and NAT function is UPnP enabled, an authorized SIP client also UPnP enabled can automatically manage NAT and firewall traversal.

In the future, proponents of UPnP see it playing a key role in home network configurations when multiple intelligent IP devices are connected in a home or small office. However, the protocol is unlikey to be used in medium and large enterprises.

UPnP references a number of IETF standards such as DHCP, HTTP Multicast UDP (HTTPMU), and Auto IP. It also uses W3C protocols such as XML and SOAP.

References

[1]   Handley, M., and V. Jacobson, "SDP: Session Description Protocol," RFC 2327, 1998.

[2]   Handley, M., C. Perkins, and E. Whelan, "Session Announcement Protocol," RFC 2974, 2000.

[3]   Mills, D., "Network Time Protocol (Version 3): Specification, Implementation, and Analysis," RFC 1305, 1992.

[4]   Schulzrinne, H., et al., "RTP: A Transport Protocol for Real-Time Applications," RFC 3550, 2003.

[5]   Schulzrinne, H., "RTP Profile for Audio and Video Conferences with Minimal Control," RFC 3550, 2003.

[6]   Duric, A., and S. Anderson, "RTP Payload Format for iLBC Speech," IETF Internet-Draft, Work in Progress, March 2003.

[7]   Rosenberg, J., and H. Schulzrinne, "An Offer/Answer Model with Session Description Protocol (SDP)," RFC 3264, 2002.

[8]   Johnston, A., and R. Sparks, "SDP Offer Answer Examples," IETF Internet-Draft, Work in Progress, June 2003.

[9]   Kent, S., and R. Atkinson, "Security Architecture for the Internet Protocol," RFC 2401, 1998.

[10] Anttalainen, T., *Introduction to Telecommunications Network Engineering*, Norwood, MA, Artech House, 1999.

[11] Schaphorst, R., *Videoconferencing and Videotelephony: Technology and Standards*, 2nd ed., Norwood, MA: Artech House, 1999.

[12] Sjoberg, J., et al., "Real-Time Transport Protocol Payload Format and File Storage Format for the Adaptive Multi-Rate (AMR) and Adaptive Multi-Rate Wideband (AMR-WB) Audio Codecs," RFC 3267, June 2002.

[13] Vermuri, A., and J. Peterson, "Session Initiation Protocol for Telephones (SIP-T): Context and Architectures," RFC 3372, 2002.

[14] Camarillo, G., et al., "Integrated Services Digital Network (ISDN) User Part (ISUP) to Session Initiation Protocol (SIP) Mapping," RFC 3398, 2002.

[15] Elwell, J., et al., "Interworking between SIP and QSIG," IETF Internet-Draft, Work in Progress, April 2003.

[16] Johnston, A., et al., "Session Initiation Protocol PSTN Call Flows," IETF Internet-Draft, Work in Progress, April 2003.

[17] Zimmerer, E., et al., "MIME Media Types for ISUP and QSIG Objects," RFC 3204, December 2001.

[18] UPnP specifications are available at http://www.upnp.org.

[19] Steinfeld, E. F., "Devices That Play Together, Work Together," *EDN Magazine*, September 2001, pp. 65–70.

8

Comparison to H.323

This chapter compares SIP to another IP telephony signaling protocol: the ITU recommendation H.323, entitled "Packet-Based Multimedia Communication." H.323 is introduced and explained using a simple call flow example. H.323 and SIP are then compared.

8.1 Introduction to H.323

H.323 [1] is an umbrella recommendation that covers all aspects of multimedia communication over packet networks. It is part of the H.32x series[1] of protocols that describes multimedia communication over ISDN, broadband (ATM),[2] telephone (PSTN), and packet (IP) networks, as shown in Table 8.1. Originally developed for video conferencing over a single LAN segment, the protocol has been extended to cover the general problem of telephony over the Internet. The first version was approved by the ITU in 1996 and was adopted by early IP telephony networks because there were no other standards. Version 2 was adopted in 1998 to fix some of the problems and limitations in version 1. Version 3 was adopted in 1999 and includes modifications and extensions to enable

---

1. In this chapter, the use of an x instead of a digit does not imply that all digits (0–9) in the range are included. In this case H.32x does not include H.325 to H.329, which have yet to be defined.

2. In this context, broadband means transported over an asynchronous transfer mode network. In anticipation of the universal deployment of ATM networks by carriers, the ITU developed a suite of protocols to support conventional telephony over ATM networks. For example, Q.2931 is the extension of Q.931 ISDN over ATM. Today, the term is used to mean high bandwidth connections—faster than modem speeds.

Table 8.1
ITU H.32x Family of Standards

Protocol	Title
H.320	Communication over ISDN networks
H.321	Communication over broadband ISDN (ATM) networks
H.322	Communication over LANs with guaranteed QoS
H.323	Communication over LANs with nonguaranteed QoS (IP)
H.324	Communication over PSTN (V.34 modems)

communications over a larger network. Version 4 was adopted in 2000 with some major changes to the protocol. Version 5 is currently under revision at the ITU-T. Currently, most deployed systems use Version 2. H.323 has been designed to be backward compatible, so, for example, a version 1 compliant end point can communicate with a version 3 gatekeeper and a version 4 end point.

H.323 references a number of other ITU and IETF protocols to completely specify the environment. Each element of the network is defined and standardized. Figure 8.1 shows the main elements: terminals, gatekeepers, gateways, and multipoint control units (MCUs). Terminals, gateways, and MCUs are network end-devices, often called end points. An end point originates and terminates media streams that could be audio, video, or data, or a combination

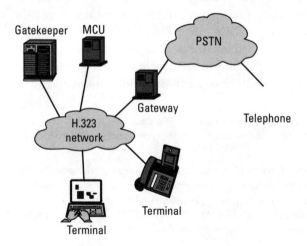

Figure 8.1 Elements of an H.323 network.

of all three. At a minimum, all H.323 end points must support basic G.711 PCM audio transmission. Support of video and data are optional. An H.323 gatekeeper is a server that controls a zone, which is the smallest administrative domain in H.323. If a gatekeeper is present, all end points within that zone must register with and defer to the gatekeeper on authorization decisions to place or accept a call. A gatekeeper also provides services to terminals in a zone, such as gateway location, address translation, bandwidth management, feature implementation, and registration. A gatekeeper is not a required element in an H.323 network, but a terminal's capabilities without one are severely limited. A gateway is another optional element in an H.323 network. It interfaces the H.323 network with another protocol network, such as the PSTN. An MCU provides conferencing services for terminals.

Some of the protocols referenced by H.323 are shown in Table 8.2. H.225 is used for registration, admission, and status (RAS), which is used for terminal-to-gatekeeper communication. A modified subset of Q.931 is used for call setup signaling between terminals. (The H.323 usage of Q.931 is not compatible with Q.931 as used in an ISDN network.) H.245 is used for control signaling or media negotiation and capability exchange between terminals. T.120 is used for multipoint graphic communications. H.323 audio codecs are specified in the ITU G.7xx series. Video codecs are specified in the H.26x series. H.323 also references two IETF protocols, RTP and RTCP, for the media transport which are described in Sections 7.2 and 7.3. The H.235 recommendation covers privacy and encryption, while H.450 covers supplementary services such as those commonly found in the PSTN (e.g., call forwarding, call hold, and call park).

Table 8.2
Protocols Referenced by H.323

Protocol	Description
H.225	Registration, admission, and status (RAS) and call signaling
H.245	Control signaling (media control)
T.120	Multipoint graphic communication
G.7xx	Audio codecs
H.26x	Video codecs
RTP	Real-time transport protocol (RFC 3550)
RTCP	RTP control protocol (RFC 3550)
H.235	Privacy and encryption
H.450	Supplementary services

8.2   Example of H.323

Figure 8.2 shows a basic call flow involving two terminals and a gatekeeper. The flow shows the interaction between the various elements and the various protocols used to establish the session. The call begins with an exchange of H.225.0 RAS messages between the calling terminal and the gatekeeper. All RAS messages are transported using UDP. It is assumed that both terminals have already registered with the gatekeeper using the Registration Request (RRQ) message. The calling terminal sends an Admission Request (ARQ) message to the gatekeeper containing the address of the called terminal and the type of session desired. The address could be specified as an H.323 alias, E.164 telephone number, e-mail address, or a URL. The gatekeeper knows about all calls in the zone it controls; it decides if the user is authorized to make a call and if there is enough bandwidth or other resources available. In this example, there is enough bandwidth, so the

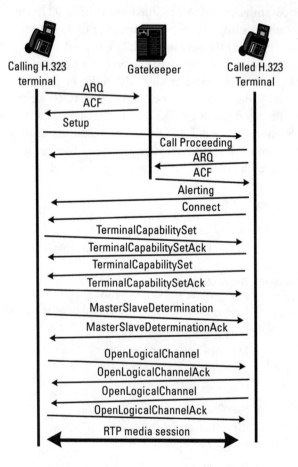

Figure 8.2 H.323 call flow example.

gatekeeper allows the call to continue by sending an Admission Confirmation (ACF) message. The ACF indicates to the calling terminal that end-point message routing, or the direct exchange of H.225 call signaling messages with the called terminal, is to be used. Alternatively, the gatekeeper can require gatekeeper routed signaling, where the gatekeeper acts like a proxy and forwards all signaling messages between the terminals. The gatekeeper has also translated the destination in the ARQ into a transport address that was returned in the ACF.

The calling terminal is now able to open a TCP connection to the called terminal using the transport address returned in the ACF and send a Q.931 Setup message to the called terminal. The called terminal responds with a Call Proceeding response to the calling terminal. The called terminal must also get permission from the gatekeeper before it accepts the call, so an ARQ is sent to the gatekeeper. When it receives the ACF from the gatekeeper, the called terminal begins alerting the user, and sends an Alerting message to the calling terminal. When the user at the calling terminal answers, a Connect message is sent. There is no acknowledgment of messages because all these messages are sent using TCP, which provides reliable transport. These call signaling messages used in H.323 are a subset of the Q.931 recommendation that covers ISDN D-channel signaling.

The next stage is the connection negotiation, which is handled by H.245 control signaling messages. A second TCP connection between the two terminals is opened by the calling terminal using the port number selected by the called terminal and returned in the Connect message. The TerminalCapabilitySet message sent contains the media capabilities of the calling terminal, listing supported codecs. It is acknowledged with a TerminalCapabilitySetAck response from the called terminal. The called terminal also sends a TerminalCapabilitySet message containing its media capabilities, which receives a TerminalCapabilitySetAck response.

The H.323 protocol requires that one terminal be selected as the master with the other as the slave. This is accomplished using MasterSlaveDetermination messages exchanged between the terminals. The messages contain the terminal type of the terminal and a random number. Terminal types are hierarchical, which determines the master. If the terminal type is the same, the random number determines the master. The message is acknowledged with a MasterSlaveDeterminationAck message.

The final phase of the call setup is the opening of two logical channels between the terminals. These channels are used to set up and control the media channels. The H.245 OpenLogicalChannel message sent in the H.245 control signaling connection contains the desired media type, including the codec that has been determined from the exchange of capabilities. It also contains a pair of addresses for the RTP and RTCP streams. The message is acknowledged with an OpenLogicalChannelAck message.

Now, the terminals begin sending RTP media packets and also RTCP control packets using the IP addresses and port numbers exchanged in the OpenLogicalChannel messages.

Figure 8.3 shows a call tear-down sequence, which either terminal may initiate. In this example, the called terminal sends an EndSessionCommand message in the H.245 control signaling channel. The other terminal responds with an EndSessionCommand message in the H.245 control signaling channel, which can now be closed. The called terminal then sends a Disengage Request (DRQ) message and receives a Disengage Confirmation (DCF) message from the gatekeeper. This way, the gatekeeper knows that the resources used in the call have now been freed up. A Call Detail Record (CDR) or other billing record can be written and stored by the gatekeeper. Next, a Q.931 Release Complete message is sent in the call signaling connection, which can then be closed. Finally, the other terminal sends a DRQ to the gatekeeper over UDP and receives a DCF response.

The call flows in Figures 8.2 and 8.3 show direct end-point signaling, where the calling terminal opens TCP connections to the called terminal and

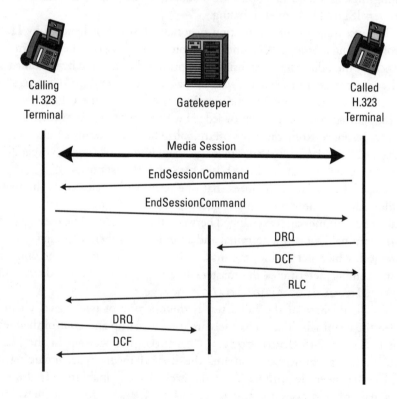

Figure 8.3   H.323 call tear-down sequence.

exchanges H.225.0 and H.245 messages. In the ACF response to the calling terminal, the gatekeeper can require gatekeeper routed signaling, where the call signaling and control signaling channels are opened with the gatekeeper, who then opens the channels with the called terminal. In this way, the gatekeeper stays in the signaling path and proxies all signaling messages. This allows the gatekeeper to know the exact call state and be able to invoke features.

8.3 Versions

There are four versions of H.323, which reflect the evolution of this protocol with a fifth underway in 2003. H.323 is fully backward compatible, so gatekeepers and terminals must support flows and mechanisms defined in all previous versions. Version 1 was approved in 1996 and was titled "Visual Telephone Systems over Networks with Non-Guaranteed Quality of Service." The example call flow in Figure 8.2 shows the version 1 call setup. Not unexpectedly given the number of messages and TCP connections, this process was very slow, sometimes taking as many as seven round trips to establish a call. While this may have been acceptable for a protocol designed for video conferencing over a single LAN segment, it is not acceptable for an IP telephony network designed to provide a similar level of service to the PSTN.

Version 2 included alternative call setup schemes to speed up the call setup. Two schemes were added to H.323, called FastStart and H.245 tunneling. FastStart is shown in Figure 8.4, in which the Setup message contains the TerminalCapabilitySet information. This saves multiple messages and round trips. In H.245 tunneling, a separate H.245 control channel is not opened. Instead, H.245 messages are encapsulated in Q.931 messages in the call signaling channel. This saves overhead in opening and closing a second TCP connection.

Versions 3 and 4 added more features to H.323 and additional annexes. Of interest to Internet devices is the support for H.323 URLs [2], full UDP support instead of TCP, and also the standardization of the use of DNS by H.323 in Annex O. Version 5 is currently being worked on by the ITU-T and is scheduled to be finished in late 2003.

8.4 Comparison

SIP and H.323 were developed for different purposes by standards bodies with very different requirements. H.323 was developed by the ITU. Its design and implementation reflects its PSTN background and heritage, utilizing binary encoding and reusing parts of ISDN signaling. SIP, on the other hand, was

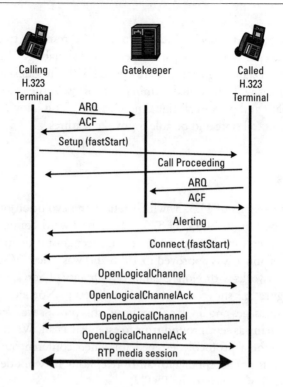

Figure 8.4 FastStart Connections with H.323.

developed by the IETF with an Internet perspective, designed to be scalable over the Internet and work in an interdomain way utilizing the full set of Internet utilities and functions.

While H.323 was deployed in early VoIP and IP videoconferencing applications, SIP with its Internet architecture is gaining momentum and is emerging as the future signaling standard for IP communications, as IP telephony is sometimes called.

8.4.1 Fundamental Differences

The first key difference is in the encoding scheme used by the protocol. SIP is a text-based protocol like HTTP and SMTP, while H.323 uses binary-encoded ASN.1 messages. The binary encoding of H.323 may result in smaller message sizes but it adds complexity to implementations. A text-based protocol such as SIP can be easily scripted and requires no tools to monitor and interpret messages—a simple packet dump from a LAN provides the ASCII-encoded SIP messages, which can be easily logged, examined, and inspected. Simple applications can be written to test and simulate SIP traffic. The text-based encoding of

SIP has made it seem more friendly to Internet and Web developers who have embraced it in developing various applications.

Another important difference is that while H.323 is exclusively a signaling protocol, SIP has both presence and instant message capability. The combination of presence and signaling in one protocol using a common universal addressing scheme, URIs, will be a very powerful driver in new applications in the future. This makes SIP an extremely powerful "rendezvous" protocol that allows a user with multiple mobile end points to locate and communicate with another user with multiple capabilities. It is for this reason that the most exciting and innovative services are being developed with SIP rather than H.323.

Another important difference is in key vendor support and momentum. SIP has been adopted by some key players in both the PC and telecommunications industries. It is widely regarded in the industry as *the* signaling protocol for session establishment over IP. While it may not immediately replace many established H.323 networks used for basic services such as simple phone calls, it is being adopted or investigated by all major players in the industry, even those who currently have a vested interest in legacy H.323 systems.

SIP has also been adopted by mobile operators as the call signaling and instant message protocol for their third generation networks currently under development. This offers the promise of millions of SIP-enabled wireless devices in the next few years. SIP is also being coupled with 802.11 wireless networks for another set of mobile services.

Another important difference is the level of security in the protocol. SIP as defined in RFC 3261 has very robust security mechanisms to provide encryption, authentication using certificates, and end-to-end message integrity even in the presence of untrusted intermediary servers. SIP did not need to develop these security features; instead, since it is an Internet protocol, it was able to inherit the rich set of Internet security protocols such as TLS and S/MIME in very straightforward way. For example, the same security mechanisms that make it safe to enter credit card information on a secure Web page form is what allows SIP to provide secure signaling between servers.

While it is not directly related to SIP, the media capability/negotiation capabilities of SDP are quite different to that provided by H.245 for H.323. H.245 is often criticized for its complexity, but SDP is often correctly criticized for its lack of expressiveness. The multicast legacy of SDP has not provided SIP with a good basis for media negotiation. As a result, some codec/media negotiation capabilities interwork poorly between implementations due to these problems. Much industry work has gone into fixing the problem, but a replacement for SDP is not yet within sight yet.

Over the years that they have coexisted, SIP and H.323 have actually become more similar in some functionality. This is natural in that each protocol seems to have adopted some of the useful features of each. For example, while SIP had DNS and URL support right from the start, H.323 began with none but has added some support for them. Another similarity relates to conferencing. While H.323 had the concept of an MCU right from the start, SIP initially had no defined conferencing mechanisms. However, work is currently underway to define the SIP signaling equivalent, called a "focus" as described in Section 11.1.2. Additional work is underway in the IETF to standardize the media mixing functions of an MCU.

In other ways, the protocols started out at opposite ends of the spectrum and have moved towards each other. For example, SIP was initially deployed exclusively with UDP, but TCP support has grown and become more important over the years. H.323 on the other hand initially could not be used over UDP exclusively, but now has been extended to be used over UDP. Another example relates to the number of messages used to set up sessions. H.323 started with a large number of messages, then reduced that number as it evolved using FastStart. Some SIP applications that use extensions such as UPDATE and PRACK to perform precondition negotiation prior to session establishment now use many more messages to set up a session than the three in the original specification. (However, this complexity in these SIP preconditions applications suggests that they, like H.323, will never achieve wide deployment, and will instead be displaced by the simpler SIP model.) Finally, H.323 started with a larger specification than SIP [3], but as SIP has grown and added functionality, it now wins the "paperweight" war in terms of more pages of specification text. As a result, most SIP versus H.323 comparisons are badly out of date due to the revisions in each protocol.

8.4.2   Strengths of Each Protocol

H.323 has carved out two niche areas in current deployed systems—it is widely deployed in small PSTN replacement networks for handling simple phone calls, and it dominates the IP videoconferencing market. Simple PSTN replacement networks that only originate and terminate phone calls without even basic features do not use most of the key advantages of SIP. As a result, they have little incentive to upgrade to SIP until the widespread adoption of SIP eventually makes SIP gateway ports much cheaper than H.323 ports. However, since SIP is earlier in its development cycle than H.323, this is not likely to happen for a number of years. New implementations of these types of systems will likely deploy SIP from the start, seeking to "future proof" the investment, but there is little incentive for deployed systems to upgrade. Also, the availability of commercial SIP to H.323 signaling gateways allows both networks to work together and complete calls [4, 5].

The videoconferencing market is dominated by PSTN ISDN devices, which have been deployed since 1990. Since H.323 shares the same heritage, it was designed to easily interwork with these existing systems. SIP was not designed to interwork with these systems and was also not designed specifically with video in mind. In fact, some key signaling components needed to do video conferencing are still not fully standardized in the IETF, although there is active work to finalize these standards. As a result, a fully functional, standards-based SIP videoconferencing system is not currently available (but will be soon). However, the deployment of videoconferencing systems has been disappointing compared to many predictions of this useful technology. This largely has to do with bandwidth and cost factors. Now that bandwidth and camera costs are falling, SIP is well positioned to handle much simpler videoconferencing systems enabled by PCs with cheap Webcams. It is likely that SIP will play a bigger role in these future software-based video systems than the current dedicated hardware videoconferencing systems. In this way, SIP may in the future be a major player in the videoconferencing market but without ever displacing the current installed base.

The major strength of SIP is that it is still relatively simple—it is an Internet protocol based on the Internet's architecture. While complex architectures have been developed using a multitude of SIP extensions, it is likely that these systems will be the exception rather than the rule. Most SIP end points will only need to support the base SIP specification and perhaps a few call control extensions to provide a wide variety of useful and innovative services.

8.5 Conclusion

This chapter has introduced H.323 and discussed and compared it to SIP. While there are some similarities between the protocols in call setup, and some niche markets that H.323 currently dominates, SIP, with its text encoding, presence and instant message extensions, and Internet architecture, is poised to be the signaling and "rendezvous" protocol of choice for Internet devices in the future.

References

[1]   "Packet-Based Multimedia Communications Systems," ITU Recommendation H.323, 2000.

[2]   Levin, O., "H.323 Uniform Resource Locator (URL) Scheme Registration," RFC 3508, 2003.

[3]   Rosenberg, J., and H. Schulzrinne, "A Comparison of SIP and H.323 for Internet Telephony," *Network and Operating System Support for Digital Audio and Video (NOSSDAV),* Cambridge, England, July 1998.

[4]   Schulzrinne, H., and C. Agboh, "Session Initiation Protocol (SIP)-H.323 Interworking Requirements," IETF Internet-Draft, Work in Progress, 2003.

[5]   Agrawl, C., "Session Initiation Protocol (SIP) – H.323 Interworking," IETF Internet-Draft, Work in Progress, 2000.

9

Wireless and 3GPP

The mobility features of SIP have been discussed in earlier chapters. In this chapter, those aspects will be explored further. In 2000 the Third Generation Partnership Project [1] adopted SIP as their call signaling protocol starting with Intelligent Multimedia core Subsystem (IMS) Release 5. Since then, a number of extensions and usage documents have been authored describing their planned use of SIP; these will be discussed in this chapter. The future direction of wireless SIP will be discussed in the final section of this chapter.

9.1 IP Mobility

There are a number of different types of mobility that will be discussed in this chapter, which include terminal mobility, personal mobility, and service mobility [2]. Terminal mobility is the ability of an end device to maintain its connection to the Internet as it moves around and possibly changes its point of connection. Personal mobility is the ability to have a constant address (identifier) across a number of devices. Finally, service mobility is the ability of a user to keep the same services when mobile.

Terminal mobility can be addressed by Mobile IP [3], which has been standardized in the IETF. It allows a terminal to keep the same IP address when roaming as it does in its home network. While roaming, the terminal is reachable by a "care of" address, which is registered in the home network. Packets destined for the roaming terminal are received in the home network, then tunneled to the terminal at the "care of" address, as shown in Figure 9.1. Mobile IP has the advantage of hiding the mobile nature of the terminal from layer 3 protocols and above. These protocols can then be used without any modification.

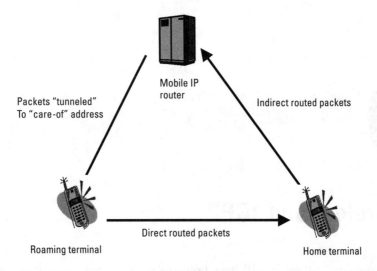

Figure 9.1 Triangular routing of IP packets in Mobile IP.

For example, a TCP connection can be maintained since the terminal appears to have a constant IP address. However, Mobile IP has the disadvantage that incoming packets are not routed directly, and as a result, most efficiently. This results in increased packet latency. While this is not a problem for nonreal-time services such as Web browsing or e-mail, real-time media transport has critical requirements on packet latency. A solution that has been proposed [2] uses the fact that a protocol such as SIP already has mobility support built in. In addition, SIP is capable of handling some of the terminal mobility aspects at the application layer. This can result in more efficient RTP packet routing and better efficiency (as the additional overhead of IP packet encapsulation required by Mobile IP are avoided).

The result is that mobile SIP devices are utilizing two different architectures. One is based on the use of Mobile IP and the other utilizes the built-in mobility support in SIP. The next sections will discuss these two approaches. As discussed in the next section, SIP is ideally suited to provide both personal and service mobility.

9.2 SIP Mobility

Personal mobility is the ability to have a constant identifier across a number of devices. A sip or sips URI has exactly this property and is fundamentally supported by SIP. SIP can also support service mobility, the ability of a user to keep the same services when mobile, although some conventions and extensions have been proposed that provide this in certain architectures.

Basic personal mobility is supported by SIP using the REGISTER method, which allows a mobile device to change its IP address and point of connection to the Internet and still be able to receive incoming calls. As discussed in Chapters 2 and 4, registration in SIP temporarily binds a user's AOR URI with a Contact URI of a particular device. As a device's IP address changes, registration allows this information to be automatically updated in the SIP network. An end device can also move between service providers using multiple layers of registrations, in which a registration is actually performed with a Contact as an address of record with another service provider. For example, consider the user agent in Figure 9.2, which has temporarily acquired a new SIP URI with a new service provider. (The reasons for doing so could include security, local policy, NAT/firewall traversal.) The user agent then performs a double registration as shown in Figure 9.2. The first registration is with the new service provider, which binds the Contact URI of the device with the new service provider's AOR URI. The second REGISTER request is routed back to the original service provider and provides the new service provider's AOR as the Contact URI. As shown later in the call flow, when a request comes in to the original service provider's network, the INVITE is redirected to the new service provider who then routes the call to the user.

For the first registration message containing the device URI would be:

```
REGISTER sip:registrar.capetown.org SIP/2.0
Via: SIP/2.0/TLS 128.5.2.1:5060;branch=z9hG4bK382112
Max-Forwards: 70
To: Nathaniel Bowditch <sip:bowditch321@capetown.org>
From: Nathaniel Bowditch <sip:bowditch321@capetown.org>
 ;tag=887865
Call-ID: 54-34-19-87-34-ar-gr
CSeq: 3 REGISTER
Contact: <sip:nat@128.5.2.1>
Content-Length: 0
```

and the second registration message with the roaming URI would be:

```
REGISTER sip:registrar.salem.ma.us SIP/2.0
Via: SIP/2.0/TLS 128.5.2.1:5060;branch=z9hG4bK1834
Max-Forwards: 70
To: Nathaniel Bowditch <sip:n.bowditch@salem.ma.us>
From: Nathaniel Bowditch <sip:n.bowditch@salem.ma.us>;tag=344231
Call-ID: 152-45-N-32-23-W3-45-43-12
CSeq: 6421 REGISTER
Contact: <sip:bowditch321@capetown.org>
Content-Length: 0
```

The first INVITE that is depicted in Figure 9.2 would be sent to sip:n.bowditch@salem.ma.us; the second INVITE would be sent to sip:bowditch321@capetown.org; and the third INVITE would be

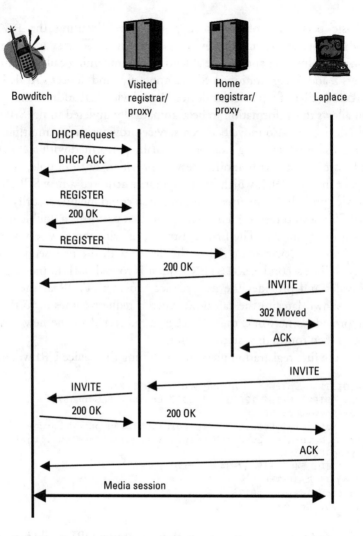

Figure 9.2 Precall mobility using SIP `REGISTER`.

sent to `sip:nat@128.5.2.1`, which reaches Bowditch and allows the session to be established. Both registrations would need to be periodically refreshed.

A disadvantage of this approach that SIP does not currently have a means to obtain a local URI. This would have to be done using a non-SIP method such as a Web page sign-up, which would be coupled with the proper authentication, authorization, and accounting mechanisms.

An optimization of this is for the local registrar to forward the registration information on the roaming user agent back to the home registrar. This has been proposed in the IETF [4] but has yet to be adopted or standardized. No changes to SIP messages are required, just a convention adopted by registrars to recognize a roaming registration and take the appropriate action. It is possible that these conventions may become standardized if the authentication and accounting systems needed to properly process such registrations are standardized in the future.

During a session, a mobile device may also change IP address as it switches between one wireless network and another (the Mobile IP protocol is not assumed—it will be discussed in the next section). Basic SIP supports this scenario as well, since a re-INVITE in a dialog can be used to update the Contact URI and change media information in the SDP. This is shown in the call flow of Figure 9.3. Here, Bowditch detects a new wireless network, uses DHCP to acquire a new IP address, then performs a re-INVITE to make the signaling and media flow to the new IP address. If the user agent momentarily is able to

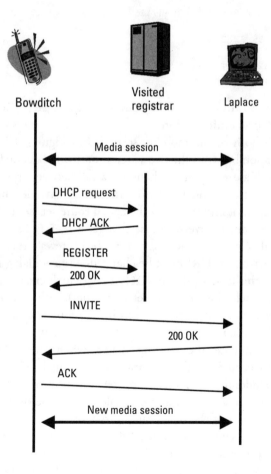

Figure 9.3 Midcall mobility using a re-INVITE.

receive media from both networks, the interruption can be almost negligible. If this is not the case, a few RTP packets may be lost as the media catches up with the signaling, resulting in a slight interruption to the call. The re-INVITE would appear as follows:

```
INVITE sip:laplace@client.mathematica.org SIP/2.0
Via: SIP/2.0/UDP 65.32.21.2:5060;branch=z9hG4bK34213
Max-Forwards: 70
To: Marquis de Laplace <sip:laplace@mathematica.org>;tag=90210
From: Nathaniel Bowditch <sip:n.bowditch@salem.ma.us>;tag=4552345
Call-ID: 413830e41eoi34ed4223123343ed21
CSeq: 5 INVITE
Contact: <sip:nat@65.43.21.2>
Content-Type: application/sdp
Content-Length: 143

v=0
o=bowditch 2590844326 2590944533 IN IP4 65.32.21.2
s=Bearing
c=IN IP4 65.32.21.2
t=0 0
m=audio 32852 RTP/AVP 0
a=rtpmap:0 PCMU/8000
```

which contains Bowditch's new IP address in the Via, Contact header fields and SDP media information.

Note that both of the mobility scenarios in Figures 9.2 and 9.3 do not require cooperation between the two wireless networks. As such, this is a useful scenario in which a user agent can hand off a call between, for example, a commercial wireless network and a home or office 802.11 wireless network.

For midcall mobility in which the actual route set (set of SIP proxies that the SIP messages must traverse) must change, a re-INVITE cannot be used. For example, if a proxy is necessary for NAT/firewall traversal, then more than just the Contact URI must be changed—a new dialog must be created. The solution to this is to send a new INVITE (which creates a new dialog and a new route set including the new set of proxies) with a Replaces header (Section 6.2.23), which identifies the existing session. The call flow is shown in Figure 9.4. It is similar to that in Figure 9.3 except that a BYE is automatically generated to terminate the existing dialog when the INVITE with the Replaces is accepted. In this scenario, the existing dialog between Bowditch and Laplace includes the old visited proxy server (this proxy Record-Routed during the initial INVITE). The new dialog using the new wireless network requires the inclusion of the new visited proxy server. As a result, an INVITE with Replaces is sent by Bowditch, which creates a new dialog that includes the new visited proxy server (which Record-Routes) but not the old visited proxy server. When Laplace accepts the INVITE, a BYE is automatically sent to terminate the old dialog that routes through the old visited proxy server that is

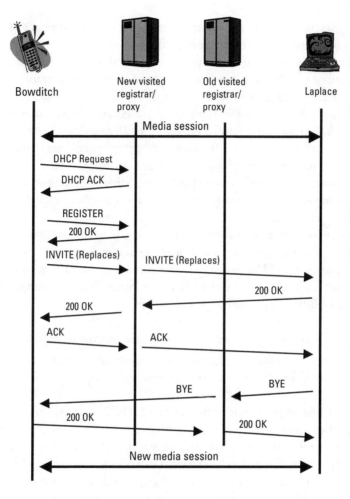

Figure 9.4 Midcall mobility using INVITE with Replaces.

now no longer involved in the session. The resulting media session is established using Bowditch's new IP address from the SDP in the INVITE.

Services in SIP can be provided in either proxies or in user agents. If the service is resident in the user agent, then there are no service mobility problems as the user moves around. However, combining service mobility and personal mobility can be challenging unless each of the user's devices are identically configured with the same services. Also, end-point resident services are only available when the end point is connected to the Internet. A terminating service such as a call forwarding service implemented in an end point will fail if the end point has temporarily lost its Internet connection. For this and other reasons, some services are implemented in the network using SIP proxy servers. For these services, service mobility for a user agent means that the same set of proxies are used to route incoming and outgoing requests when mobile.

Due to the nature of the Internet, in general, there is no reason why a user agent cannot use the same proxies when connected to the Internet at a different point. That is, a user agent that is normally in the United States that is configured to use a set of proxies in the United States can still use those proxies when roaming in Europe, for example. Perhaps the SIP hops will have a slightly higher latency due to more router hops and a call setup request may take a second or two longer to complete. However, this has no impact on the quality of the media session as the media always flows directly between the two user agents and does not traverse the SIP proxy servers. As a result, SIP can easily support service mobility over the Internet.

However, there are some cases in which this service mobility approach will not work. For example, if a local proxy server must be traversed in order to facilitate firewall or NAT traversal, or for some other security reason, then a user agent may have to use a different first hop proxy when roaming. In this case, service mobility is still possible provided that:

1. The roaming user agent is able to discover the necessary local proxy.
2. Both incoming and outgoing requests are routed through the home proxy in addition to any local proxies.

The first requirement is met by the DHCP extension to SIP [5], which allows a user agent to learn of a local proxy server at the same time it learns its IP address and other IP configuration information. The second requirement is met using a "preloaded" Route header field in requests. Normally a Route header is inserted in a request when a proxy requests it using a Record-Route header field. However, it is possible for a configured user agent to include a Route header field. If the Route header contains the URI of the home proxy, the request will be routed to the home proxy after the local proxies have been traversed, meeting the requirement for outgoing requests. For incoming requests, the double registration technique will result in both the home and local proxies being traversed by incoming requests. This will result in a call flow similar to Figure 9.2 but with the home proxy server forwarding the INVITE instead of redirecting.

These SIP mobility capabilities are well suited to use over a wireless network such as 802.11 in a home, office, or public space. As roaming agreements allow such wireless "hotspots" to be linked up in metropolitan areas, this will provide a wireless service. However, commercial wireless providers plan a specific purpose wireless telephony network using SIP. For some of their business and service requirements, some SIP extensions have been developed, which will be discussed in this chapter.

Wireless SIP clients may also make use of voice codecs such as the iLBC [6], which is highly tolerant to packet loss, which may be experienced in a heavily loaded 802.11 network.

9.3 3GPP Architecture and SIP

The 3GPP architecture uses SIP in the IP Multimedia Core Network Subsystem (IMS). A simplified architecture is shown in Figure 9.5. The elements of the IMS architecture are listed in Table 9.1. The specification of SIP in 3GPP is described in 3GPP TS 24.229 and in [7].

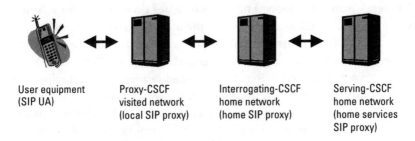

User equipment	Proxy-CSCF	Interrogating-CSCF	Serving-CSCF
(SIP UA)	visited network	home network	home network
	(local SIP proxy)	(home SIP proxy)	(home services SIP proxy)

Figure 9.5 3GPP IMS architecture.

Table 9.1
IMS Elements

Abbreviation	Name
P-CSCF	Proxy – Call Session Control Function
I-CSCF	Interrogating – Call Session Control Function
S-CSCF	Serving – Call Session Control Function
UE	User Equipment
MGCF	Media Gateway Control Function
MGW	Media Gateway
AS	Application Server
MRFC	Media Resource Function Controller
BGCF	Breakout Gateway Control Function
HSS	Home Subscriber Server

The 3GPP architecture relies on Mobile IP instead of some of the mobility aspects of SIP described in the previous section. The reasons for doing so are primarily business related rather then technical in nature.

Another requirement of mobility systems is a keep alive signal, which allows end points and proxies to know that a user agent still has network connectivity. On an end point to end point basis, this can be done using RTCP (see Section 7.2) reports sent periodically, even when the media is on hold or silence suppression is taking place. However, proxies do not have access to these direct end-to-end reports. Instead, the Session Timer extension [8] and re-INVITEs can be used for this purpose.

CSCF are SIP proxies that also sometimes behave as a B2BUA under certain circumstances. For example, if a P-CSCF looses the radio link to the UE (user equipment) that contains the SIP UA, it can send a BYE on behalf of the UE to tear down the session. The motivation for doing this is for sessions that have a per-minute billing charge, which the out-of-contact UE would otherwise have to pay for but not have the ability to disconnect. To save bandwidth on the wireless connection, a P-CSCF removes Route, Record-Route, Path, Via, Service-Route and other header fields and reinserts them in the opposite direction. To prevent high bandwidth codecs from being used by a UE, a P-CSCF may edit the list of codecs in an SDP offer or answer, preventing the codec from being used. A P-CSCF may change the To and From headers to provide privacy, which is a B2BUA function.

The Proxy CSCF provides emergency service, triggers for local services, and does telephone number normalization for the rest of the network. The P-CSCF is used as the default outbound proxy server for a UE outside its home network. The Interrogating CSCF queries the HSS to determine the proper Service CSCF. The I-CSCF also can do hiding of the S-CSCF network by removing or encrypting Via header fields. The Serving CSCF provides the services for the subscriber. It identifies the user's service profile and privileges.

The 3GPP plans to exclusively use IPv6 addresses. This is due to the number of 3GPP subscribers envisioned and the fact that with Mobile IP, each device may use more than one IP address at a time. SIP RFC 3261 includes full support for IPv6 addresses, and an extension to SDP [9] adds IPv6 support to SDP.

The 3GPP also uses signaling compression [10] to co mpress SIP messages transmitted over a wireless link. This is primarily done to minimize latency rather than bandwidth savings. The use of signaling compression with SIP is described in [11], which defines a parameter comp=sigcomp that can be used in Via header fields and as a URI parameter that can be used, for example, in a Contact header field.

The 3GPP uses the adaptive multirate (AMR) [12] codec for the audio encoding.

9.4   3GPP Header Fields

Some SIP header fields have been developed based on 3GPP requirements and are discussed in the next sections. They are listed here instead of the main listing of SIP header fields in Chapter 6 since their use is not well standardized outside the 3GPP architecture. In the future, however, their use may be standardized within the IETF and they may be used more widely.

9.4.1   Service-Route

The Service-Route header field [13] can be used in a 2xx response to a REGISTER request. It can be used by a registrar server to provide to the registering UA URIs to include in a preloaded Route header field in future requests. The Service-Route URIs are only valid for the duration of the registration and should be updated when the registration is refreshed.

An example is:

```
Service-Route: <sip:proxy23.service.provider.com;lr>
```

9.4.2   Path

The Path header field [14] is an optional header field in REGISTER requests. It can be thought of as a Record-Route mechanism for REGISTER requests, which establishes a route set that is valid for the duration of the registration. The Path header field may be inserted by a proxy, which forwards a REGISTER request to a registrar server. The registrar copies the Path header field into the 200 OK response to the REGISTER, which then provides the route set information to the user agent that is registering. In a mobile network, the Path header field can be used to discover and inform the user agent of the proxies can be used to populate preloaded Route header fields.

An example is:

```
Path: <sip:proxy2.another.provider.com;lr>
```

9.4.3   Other P-Headers

In addition, some headers that are specific to 3GPP have been defined [15]. These so-called P-headers (which stands for proprietary, preliminary, or private) are defined in syntax only in an informational RFC per the SIP change process [16]. Examples of the use of these P-headers are given in the call flow of Section 10.7. They are listed in Table 9.2.

Table 9.2
3GPP P-Headers

Header Field	Use
P-Associated-URI	Lists other URIs associated with the user
P-Called-Party-ID	Lists the URI of the called party
P-Visited-Network-ID	Identifies the visited network
P-Access-Network-Info	Identifies the access network
P-Charging-Function-Addresses	Contains charging information
P-Charging-Vector	More charging information

Source: [15].

9.5 Future of SIP and Wireless

It is clear that as IP networks become increasingly wireless, SIP will often be utilized over wireless networks. It is well suited for such use for the reasons discussed in this chapter: it has both built-in mobility support when Mobile IP is not used, and can also be used with Mobile IP depending on the wireless network design.

Additional work on authentication and roaming will likely be done with SIP as the extensions developed for the 3GPP architecture are too specific to be useful in most other networks.

References

[1]   Information about the 3GPP project, including the latest technical specifications (TS), can be found at http://www.3gpp.org.

[2]   Schulzrinne, H., and E. Wedlund, "Application-Layer Mobility Using SIP," *Mobility Mobile Computing and Communications Review (MC2R)*, Vol. 4, No. 3, July 2000.

[3]   Perkins, C., "IP Mobility Support," RFC 2002, 1996.

[4]   Vakil, F., et al., "Supporting Mobility for Multimedia with SIP," IETF Internet-Draft, Work in Progress, December 2000.

[5]   Schulzrinne, H., "Dynamic Host Configuration Protocol (DHCP-for-IPv4) Option for Session Initiation Protocol (SIP) Servers," RFC 3361, 2002.

[6]   Duric, A., and S. Anderson, "RTP Payload Format for iLBC Speech," IETF Internet-Draft, Work in Progress, March 2003.

[7]   3GPP TS relating to SIP include TS 23.228 and TS 24.229. For the latest on these and other SIP-related specifications, visit http://www.3gpp.org.

[8]   Donovan, S., and J. Rosenberg, "Session Timers in the Session Initiation Protocol (SIP)," IETF Internet-Draft, Work in Progress, November 2002.

[9]   Olson, S., G. Camarillo, and A. Roach, "Support for IPv6 in Session Description Protocol (SDP)," RFC 3266, 2002.

[10] Price, R., et al., "Signaling Compression (SigComp)," RFC 3320, 2003.

[11] Camarillo, G., "Compressing the Session Initiation Protocol (SIP)," RFC 3486, February 2003.

[12] Sjoberg, J., et al., "Real-Time Transport Protocol Payload Format and File Storage Format for the Adaptive Multi-Rate (AMR) and Adaptive Multi-Rate Wideband (AMR-WB) Audio Codecs," RFC 3267, June 2002.

[13] Willis, D., and B. Hoeneisen, "Session Initiation Protocol Extension Header Field for Service Route Discovery During Registration," IETF Internet-Draft, Work in Progress, February 2003.

[14] Willis, D., and B. Hoeneisen, "Session Initiation Protocol (SIP) Extension Header Field for Registering Non-Adjacent Contacts," RFC 3327, 2003.

[15] Garcia-Martin, M., E. Henrikson, and D. Mills, "Private Header (P-Header) Extensions to the Session Initiation Protocol (SIP) for the 3rd-Generation Partnership Project (3GPP)," RFC 3255, 2003

[16] Mankin, A., et al., "Change Process for the Session Initiation Protocol (SIP)," RFC 3427, 2002.

10

Call Flow Examples

In this chapter, many of the concepts and details presented in the preceding chapters will be illustrated with examples. Each example includes a call flow diagram, a discussion of the example, followed by the message details. Each message is labeled in the figure with a message number for easy reference. For more examples of the protocol, refer to the SIP specification [1] and the SIP call flows [2, 3] documents.

The purpose of the examples in this chapter is to illustrate aspects of the SIP protocol. The interoperation scenarios with the PSTN and with an H.323 network are not intended to fully define the interworking or show a complete parameter mapping between the protocols. Likewise, simplifications such as minimal authentication and direct client-to-gateway messaging are used to make the examples more clear.

10.1 SIP Call with Authentication, Proxies, and Record-Route

Figure 10.1 shows a basic SIP call between two SIP user agents involving two proxy servers. Rather than perform a DNS query on the SIP URI of the called party, the calling SIP phone sends the INVITE request to a proxy server for address resolution. The proxy server requires authentication to perform this service and replies with a 407 Proxy Authorization Required response. Using the nonce from the challenge, the caller resends the INVITE with the caller's username and password credentials encrypted. The proxy checks the credentials, and finding them to be correct, performs the DNS lookup on the Request-URI. The INVITE is then forwarded to the proxy server listed in the DNS SRV record that handles the language.org

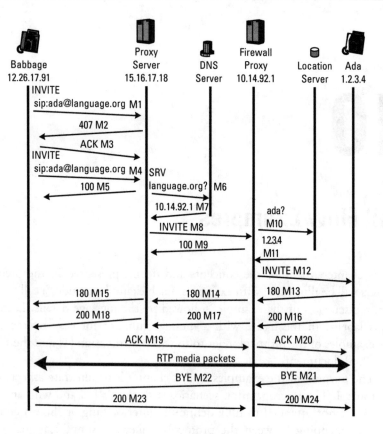

Figure 10.1 SIP-to-SIP call with authentication, proxies, and record-route.

domain. That proxy then looks up the Request-URI and locates a registration for the called party. The INVITE is forwarded to the destination UAS, a Record-Route header having been inserted to ensure that the proxy is present in all future requests by either party. This is because a direct routed SIP message to Ada would be blocked by the firewall.

The called party receives the INVITE request and sends 180 Ringing and 200 OK responses, which are routed back to the caller using the Via header chain from the initial INVITE. The ACK sent by the caller includes a Route header built from the Record-Route header field in the 200 OK response. This routing skips the first proxy but includes the firewall proxy. The media session begins with the user agents exchanging RTP and RTCP packets.

The call terminates when the called party, Ada, sends a BYE, which includes a Route header generated from the Record-Route header field in the INVITE. Note that the CSeq for the called user agent is initialized to

1000. The acknowledgment of the BYE with a 200 OK response causes both sides to stop sending media packets.

```
M1   INVITE sip:ada@language.org SIP/2.0                    ⇐Request-URI
     Via: SIP/2.0/UDP 12.26.17.91:5060;branch=z9hG4bK454
     Max-Forwards: 70
     From: Charles Babbage <sip:babbage@analyticalsoc.org>;tag=9382
     To: sip:ada@language.org
     Call-ID: f6-32-9a-34-91-e7@analyticalsoc.org
     CSeq: 1 INVITE                                        ⇐CSeq initialized to 1
     Contact: <sip:babbage@client.analyticalsoc.org>
     Subject: RE: Software
     User-Agent: Difference Engine 1
     Content-Type: application/sdp
     Content-Length: 137

     v=0
     o=Babbage 2890844534 2890844534 IN IP4 12.26.17.91
     s=-
     t=0 0
     c=IN IP4 12.26.17.91                                  ⇐Babbage's IP address
     m=audio 49170 RTP/AVP 0                               ⇐Port number
     a=rtpmap:0 PCMU/8000                                  ⇐Codec info



M2   SIP/2.0 407 Proxy Authentication Required
     Via: SIP/2.0/UDP 12.26.17.91:5060;branch=z9hG4bK454
     From: Charles Babbage <sip:babbage@analyticalsoc.org>;tag=9382
     To: <sip:ada@language.org>;tag=34q4356g
     Call-ID: f6-32-9a-34-91-e7@analyticalsoc.org
     CSeq: 1 INVITE
     Proxy-Authenticate: Digest                            ⇐Authentication
       realm="language.org",                                 challenge
       nonce="9c8e88df84f1cec4341ae6e5a359",
       opaque="", stale=FALSE, algorithm=MD5



M3   ACK sip:ada@language.org SIP/2.0
     Via: SIP/2.0/UDP 12.26.17.91:5060;branch=z9hG4bK454
     Max-Forwards: 70
     From: Charles Babbage <sip:babbage@analyticalsoc.org>;tag=9382
     To: <sip:ada@language.org>;tag=34q4356g
     Call-ID: f6-32-9a-34-91-e7@analyticalsoc.org
     CSeq: 1 ACK                                           ⇐CSeq not incremented
                                                            Method set to ACK



M4   INVITE sip:ada@language.org SIP/2.0                   ⇐INVITE resent with
     Via: SIP/2.0/UDP 12.26.17.91:5060;branch=z9hG4bK221     credentials
     Max-Forwards: 70
     From: Charles Babbage <sip:babbage@analyticalsoc.org>;tag=9382
     To: sip:ada@language.org
```

```
        Call-ID:f6-32-9a-34-91-e7@analyticalsoc.org          ⇐Call-ID unchanged
        CSeq: 2 INVITE                                        ⇐CSeq incremented
        Proxy-Authorization: Digest
         username="Babbage",                                  ⇐Credentials
         realm="language.org",
         nonce="9c8e88df84f1cec4341ae6e5a359",
         opaque="", response="e56131d19580cd833064787ecc"
        Contact: <sip:babbage@client.analyticalsoc.org>
        Subject: RE: Software
        User-Agent: Difference Engine 1
        Content-Type: application/sdp
        Content-Length: 137

        v=0
        o=Babbage 2890844534 2890844534 IN IP4 12.26.17.91
        s=-
        t=0 0
        c=IN IP4 12.26.17.91
        m=audio 49170 RTP/AVP 0
        a=rtpmap:0 PCMU/8000


M5      SIP/2.0 100 Trying                                    ⇐Credentials accepted
        Via: SIP/2.0/UDP 12.26.17.91:5060;branch=z9hG4bK221
        From: Charles Babbage <sip:babbage@analyticalsoc.org>;tag=9382
        To: sip:ada@language.org
        Call-ID: f6-32-9a-34-91-e7@analyticalsoc.org
        CSeq: 2 INVITE


M6      DNS Query: SRV language.org?


M7      DNS SRV Record: 10.14.92.1


M8      INVITE sip:ada@language.org SIP/2.0
        Via: SIP/2.0/UDP 15.16.17.18:5060;branch=z9hG4bK3f31049.1
        Via: SIP/2.0/UDP 12.26.17.91:5060;branch=z9hG4bK221
        Max-Forwards: 69
        From: Charles Babbage <sip:babbage@analyticalsoc.org>;tag=9382
        To: sip:ada@language.org
        Call-ID: f6-32-9a-34-91-e7@analyticalsoc.org
        CSeq: 2 INVITE
        Contact: <sip:babbage@client.analyticalsoc.org>
        Subject: RE: Software
        User-Agent: Difference Engine 1
        Content-Type: application/sdp
        Content-Length: 137

        v=0
        o=Babbage 2890844534 2890844534 IN IP4 12.26.17.91
        s=-
        t=0 0
```

```
      c=IN IP4 12.26.17.91
      m=audio 49170 RTP/AVP 0
      a=rtpmap:0 PCMU/8000
```

M9   SIP/2.0 100 Trying   ⇐Not Forwarded
      Via:SIP/2.0/UDP 15.16.17.18:5060;branch=z9hG4bK3f31049.1
      Via: SIP/2.0/UDP 12.26.17.91:5060;branch=z9hG4bK221
      From: Charles Babbage <sip:babbage@analyticalsoc.org>;tag=9382
      To: sip:ada@language.org
      Call-ID: f6-32-9a-34-91-e7@analyticalsoc.org
      CSeq: 2 INVITE

M10   Location Service Query: ada?

M11   Location Service Response: 1.2.3.4

M12   INVITE sip:ada@1.2.3.4 SIP/2.0
      Via: SIP/2.0/UDP 10.14.92.1:5060;branch=z9hG4bK24105.1
      Via: SIP/2.0/UDP 15.16.17.18:5060;branch=z9hG4bK3f31049.1
      Via: SIP/2.0/UDP 12.26.17.91:5060;branch=z9hG4bK221
      Max-Forwards: 68
      From: Charles Babbage <sip:babbage@analyticalsoc.org>;tag=9382
      To: sip:ada@language.org
      Call-ID: f6-32-9a-34-91-e7@analyticalsoc.org
      CSeq: 2 INVITE
      Contact: <sip:babbage@client.analyticalsoc.org>
      Subject: RE: Software
      User-Agent: Difference Engine 1
      Record-Route: <sip:10.14.92.1;lr>   ⇐Record-route added by proxy
      Content-Type: application/sdp
      Content-Length: 137

      v=0
      o=Babbage 2890844534 2890844534 IN IP4 12.26.17.91
      s=-
      t=0 0
      c=IN IP4 12.26.17.91
      m=audio 49170 RTP/AVP 0
      a=rtpmap:0 PCMU/8000

M13   SIP/2.0 180 Ringing
      Via: SIP/2.0/UDP 10.14.92.1:5060;branch=z9hG4bK24105.1
      Via: SIP/2.0/UDP 15.16.17.18:5060;branch=z9hG4bK3f31049.1
      Via: SIP/2.0/UDP 12.26.17.91:5060;branch=z9hG4bK221
      From: Charles Babbage <sip:babbage@analyticalsoc.org>;tag=9382
      To: <sip:ada@language.org>;tag=65a3547e3   ⇐Tag added by
      Call-ID: f6-32-9a-34-91-e7@analyticalsoc.org   called party
      CSeq: 2 INVITE
      Contact: sip:ada@drawingroom.language.org
```
```

```
Record-Route: <sip:10.14.92.1;lr>
```

M14 SIP/2.0 180 Ringing
```
     Via: SIP/2.0/UDP 15.16.17.18:5060;branch=z9hG4bK3f31049.1
     Via: SIP/2.0/UDP 12.26.17.91:5060;branch=z9hG4bK221
     From: Charles Babbage <sip:babbage@analyticalsoc.org>;tag=9382
     To: <sip:ada@language.org>;tag=65a3547e3
     Call-ID: f6-32-9a-34-91-e7@analyticalsoc.org
     CSeq: 2 INVITE
     Contact: sip:ada@drawingroom.language.org
     Record-Route: <sip:10.14.92.1;lr>
```

M15 SIP/2.0 180 Ringing
```
     Via: SIP/2.0/UDP 12.26.17.91:5060;branch=z9hG4bK221
     From: Charles Babbage <sip:babbage@analyticalsoc.org>;tag=9382
     To: <sip:ada@language.org>;tag=65a3547e3
     Call-ID: f6-32-9a-34-91-e7@analyticalsoc.org
     CSeq: 2 INVITE
```

M16 SIP/2.0 200 OK   ⇐Call accepted
```
     Via: SIP/2.0/UDP 10.14.92.1:5060;branch=z9hG4bK24105.1
     Via: SIP/2.0/UDP 15.16.17.18:5060;branch=z9hG4bK3f31049.1
     Via: SIP/2.0/UDP 12.26.17.91:5060;branch=z9hG4bK221
     From: Charles Babbage <sip:babbage@analyticalsoc.org>;tag=9382
     To: <sip:ada@language.org>;tag=65a3547e3
     Call-ID: f6-32-9a-34-91-e7@analyticalsoc.org
     CSeq: 2 INVITE
     Contact: sip:ada@drawingroom.language.org
     Record-Route: <sip:10.14.92.1;lr>
     Content-Type: application/sdp
     Content-Length: 126
```
```
     v=0
     o=Ada 2890844536 2890844536 IN IP4 1.2.3.4
     s=-
     t=0 0
     c=IN IP4 1.2.3.4                                   ⇐Ada'sIPaddress
     m=audio 52310 RTP/AVP 0                            ⇐Port number
     a=rtpmap:0 PCMU/8000                               ⇐Codec information
```

M17 SIP/2.0 200 OK
```
     Via: SIP/2.0/UDP 15.16.17.18:5060;branch=z9hG4bK3f31049.1
     Via: SIP/2.0/UDP 12.26.17.91:5060;branch=z9hG4bK221
     From: Charles Babbage <sip:babbage@analyticalsoc.org>;tag=9382
     To: <sip:ada@language.org>;tag=65a3547e3
     Call-ID: f6-32-9a-34-91-e7@analyticalsoc.org
     CSeq: 2 INVITE
     Contact: sip:ada@drawingroom.language.org
     Record-Route: <sip:10.14.92.1;lr>
     Content-Type: application/sdp
```

```
Content-Length: 126

v=0
o=Ada 2890844536 2890844536 IN IP4 1.2.3.4
s=-
t=0 0
c=IN IP4 1.2.3.4
m=audio 52310 RTP/AVP 0
a=rtpmap:0 PCMU/8000
```

M18   SIP/2.0 200 OK
```
       Via: SIP/2.0/UDP 12.26.17.91:5060;branch=z9hG4bK221
       From: Charles Babbage <sip:babbage@analyticalsoc.org>;tag=9382
       To: <sip:ada@language.org>;tag=65a3547e3
       Call-ID: f6-32-9a-34-91-e7@analyticalsoc.org
       CSeq: 2 INVITE
       Contact: sip:ada@drawingroom.language.org
       Record-Route: <sip:10.14.92.1;lr>
       Content-Type: application/sdp
       Content-Length: 126

       v=0
       o=Ada 2890844536 2890844536 IN IP4 1.2.3.4
       s=-
       t=0 0
       c=IN IP4 1.2.3.4
       m=audio 52310 RTP/AVP 0
       a=rtpmap:0 PCMU/8000
```

M19   ACK sip:ada@drawingroom.language.org SIP/2.0   ⇐Sent to ALG
```
       Via: SIP/2.0/UDP 12.26.17.91:5060;branch=z9hG4bK789
       Max-Forwards: 70
       From: Charles Babbage <sip:babbage@analyticalsoc.org>;tag=9382
       To: <sip:ada@language.org>;tag=65a3547e3
       Call-ID: f6-32-9a-34-91-e7@analyticalsoc.org
       CSeq: 2 ACK
       Route: <sip:10.14.92.1;lr>
```

M20   ACK sip:ada@drawingroom.language.org SIP/2.0
```
       Via: SIP/2.0/UDP 10.14.92.1:5060;branch=z9hG4bK24105.1
       Via: SIP/2.0/UDP 12.26.17.91:5060;branch=z9hG4bK789
       Max-Forwards: 69
       From: Charles Babbage <sip:babbage@analyticalsoc.org>;tag=9382
       To: <sip:ada@language.org>;tag=65a3547e3
       Call-ID: f6-32-9a-34-91-e7@analyticalsoc.org
       CSeq: 2 INVITE
```

M21   BYE sip:babbage@client.analyticalsoc.org SIP/2.0
```
       Via: SIP/2.0/UDP 1.2.3.4:5060;branch=z9hG4bK543
       Max-Forwards: 70
```

```
        From: Ada Lovelace <sip:ada@language.org>;tag=65a3547e3
        To: Charles Babbage <sip:babbage@analyticalsoc.org>;tag=9382
        Call-ID: f6-32-9a-34-91-e7@analyticalsoc.org
        CSeq: 1000 BYE                              ⇐CSeq initialized to 1000
        Route: <sip:10.14.92.1;lr>                  ⇐From Record-Route header


M22   BYE sip:babbage@client.analyticalsoc.org SIP/2.0
        Via: SIP/2.0/UDP 10.14.92.1:5060;branch=z9hG4bK24105.1
        Via: SIP/2.0/UDP 1.2.3.4:5060;branch=z9hG4bK543
        Max-Forwards: 69
        From: Ada Lovelace <sip:ada@language.org>;tag=65a3547e3
        To: Charles Babbage <sip:babbage@analyticalsoc.org>;tag=9382
        Call-ID: f6-32-9a-34-91-e7@analyticalsoc.org
        CSeq: 1000 BYE


M23   SIP/2.0 200 OK
        Via: SIP/2.0/UDP 10.14.92.1:5060;branch= z9hG4bK24105.1
        Via: SIP/2.0/UDP 1.2.3.4:5060;branch=z9hG4bK543
        From: Ada Lovelace <sip:ada@language.org>;tag=65a3547e3
        To: Charles Babbage <sip:babbage@analyticalsoc.org>;tag=9382
        Call-ID: f6-32-9a-34-91-e7@analyticalsoc.org
        CSeq: 1000 BYE


M24   SIP/2.0 200 OK
        Via: SIP/2.0/UDP 1.2.3.4:5060;branch=z9hG4bK543
        From: Ada Lovelace <sip:ada@language.org>;tag=65a3547e3
        To: Charles Babbage <sip:babbage@analyticalsoc.org>;tag=9382
        Call-ID: f6-32-9a-34-91-e7@analyticalsoc.org
        CSeq: 1000 BYE
```

10.2   SIP Call with Stateless and Stateful Proxies with Called Party Busy

Figure 10.2 shows an example of a SIP with a stateless proxy server and a stateful proxy server. The call is not completed because called party is busy. The called user agent initially sends a 180 Ringing response but then sends a 600 Busy Everywhere response containing a Retry-After header to indicate that the call is being rejected. The stateful proxy returns a 100 Trying response to the INVITE, and also acknowledges the 600 Busy Everywhere response with an ACK. The stateless proxy does not send a 100 Trying and forwards the 600 Busy Everywhere and the ACK sent by the caller user agent. Also note that the initial INVITE does not contain a message body.

```
M1    INVITE sip:schockley@transistor.org SIP/2.0
        Via: SIP/2.0/UDP discrete.sampling.org:5060;branch=z9hG4bK5654
```

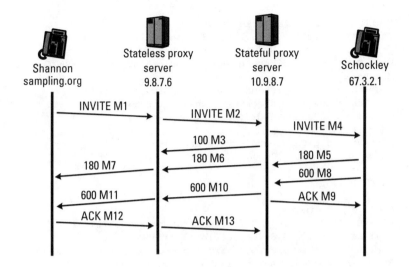

Figure 10.2 SIP call example with stateless and stateful proxies with busy called party.

```
     Max-Forwards: 70
     From: Shannon <sip:shannon@sampling.org>;tag=cgdf4
     To: Schockley <sip:shockley@transistor.com>
     Call-ID: adf8gasdd7f1d@discrete.sampling.org
     CSeq: 1 INVITE
     Contact: <sip:Shannon@discrete.sampling.org>
     Date: Sat, 8 Jul 2000 08:23:00 GMT           ⇐Optional date header
     Content-Length: 0                         ⇐Optional Content-Length header


M2   INVITE sip:schockley@transistor.org SIP/2.0           ⇐Stateless proxy
     Via: SIP/2.0/UDP 9.8.7.6:5060;branch=z9hG4bK1.1        does not send 100
     Via: SIP/2.0/UDP discrete.sampling.org:5060;branch=z9hG4bK5654
     Max-Forwards: 69
     From: Shannon <sip:shannon@sampling.org>
     To: Schockley <sip:shockley@transistor.com>
     Call-ID: adf8gasdd7f1d@discrete.sampling.org
     CSeq: 1 INVITE
     Contact: <sip:Shannon@discrete.sampling.org>
     Date: Sat, 8 Jul 2000 08:23:00 GMT
     Content-Length: 0


M3   SIP/2.0 100 Trying                              ⇐Stateful proxy does send 100
     Via: SIP/2.0/UDP 9.8.7.6:5060;branch=z9hG4bK1.1
     Via: SIP/2.0/UDP discrete.sampling.org:5060;branch=z9hG4bK5654
     From: Shannon <sip:shannon@sampling.org>;tag=cgdf4
     To: Schockley <sip:shockley@transistor.com>
     Call-ID: adf8gasdd7f1d@discrete.sampling.org
     CSeq: 1 INVITE
     Content-Length: 0
```

```
M4    INVITE sip:schockley@transistor.org SIP/2.0
      Via: SIP/2.0/UDP 10.9.8.7:52103;branch=z9hG4bkff7d.1
      Via: SIP/2.0/UDP 9.8.7.6:5060;branch=z9hG4bK1.1
      Via: SIP/2.0/UDP discrete.sampling.org:5060;branch=z9hG4bK5654
      Max-Forwards: 68
      From: Shannon <sip:shannon@sampling.org>;tag=cgdf4
      To: Schockley <sip:shockley@transistor.com>
      Call-ID: adf8gasdd7fld@discrete.sampling.org
      CSeq: 1 INVITE
      Contact: <sip:Shannon@discrete.sampling.org>
      Date: Sat, 8 Jul 2000 08:23:00 GMT
      Content-Length: 0


M5    SIP/2.0 180 Ringing
      Via: SIP/2.0/UDP 10.9.8.7:52103;branch=z9hG4bKff7d.1
      Via: SIP/2.0/UDP 9.8.7.6:5060;branch=z9hG4bK1.1
      Via: SIP/2.0/UDP discrete.sampling.org:5060;branch=z9hG4bK5654
      From: Shannon <sip:shannon@sampling.org>;tag=cgdf4
      To: Schockley <sip:shockley@transistor.com>;tag=1
      Call-ID: adf8gasdd7fld@discrete.sampling.org
      CSeq: 1 INVITE
      Contact: <sip:shockley@4.5.6.7>
      Content-Length: 0


M6    SIP/2.0 180 Ringing
      Via: SIP/2.0/UDP 9.8.7.6:5060;branch=z9hG4bK1.1
      Via: SIP/2.0/UDP discrete.sampling.org:5060;branch=z9hG4bK5654
      From: Shannon <sip:shannon@sampling.org>;tag=cgdf4
      To: Schockley <sip:shockley@transistor.com>;tag=1
      Call-ID: adf8gasdd7fld@discrete.sampling.org
      CSeq: 1 INVITE
      Contact: <sip:shockley@4.5.6.7>
      Content-Length: 0


M7    SIP/2.0 180 Ringing
      Via: SIP/2.0/UDP discrete.sampling.org:5060;branch=z9hG4bK5654
      From: Shannon <sip:shannon@sampling.org>;tag=cgdf4
      To: Schockley <sip:shockley@transistor.com>;tag=1
      Call-ID: adf8gasdd7fld@discrete.sampling.org
      CSeq: 1 INVITE
      Contact: <sip:shockley@4.5.6.7>
      Content-Length: 0


M8    SIP/2.0 600 Busy Everywhere                    ⇐Schockley is busy
      Via: SIP/2.0/UDP 10.9.8.7:52103;branch=z9hG4bKff7d.1
      Via: SIP/2.0/UDP 9.8.7.6:5060;branch=z9hG4bK1.1
      Via: SIP/2.0/UDP discrete.sampling.org:5060;branch=z9hG4bK5654
      From: Shannon <sip:shannon@sampling.org>;tag=cgdf4
      To: Schockley <sip:shockley@transistor.com>;tag=1
      Call-ID: adf8gasdd7fld@discrete.sampling.org
```

```
     CSeq: 1 INVITE
     Retry-After: Sun, 9 Jul 2000 11:59:00 GMT
     Content-Length: 0
```

M9   ACK sip:schockley@transistor.com SIP/2.0   ⇐Stateful proxy does ACK
```
     Via: SIP/2.0/UDP 10.9.8.7:52103;branch=z9hG4bK5f7e.1
     Max-Forwards: 70
     From: Shannon <sip:shannon@sampling.org>;tag=cgdf4
     To: Schockley <sip:shockley@transistor.com>;tag=1
     Call-ID: adf8gasdd7fld@discrete.sampling.org
     CSeq: 1 ACK
     Content-Length: 0
```

M10 SIP/2.0 600 Busy Everywhere
```
     Via: SIP/2.0/UDP 9.8.7.6:5060;branch=z9hG4bK1.1
     Via: SIP/2.0/UDP discrete.sampling.org:5060;branch=z9hG4bK5654
     From: Shannon <sip:shannon@sampling.org>;tag=cgdf4
     To: Schockley <sip:shockley@transistor.com>;tag=1
     Call-ID: adf8gasdd7fld@discrete.sampling.org
     CSeq: 1 INVITE
     Retry-After: Sun, 9 Jul 2000 11:59:00 GMT
     Content-Length: 0
     Call Flow Examples 163
```

M11 SIP/2.0 600 Busy Everywhere   ⇐Stateless proxy does not ACK response
```
     Via: SIP/2.0/UDP discrete.sampling.org:5060;branch=z9hG4bK5654
     From: Shannon <sip:shannon@sampling.org>;tag=cgdf4
     To: Schockley <sip:shockley@transistor.com>;tag=1
     Call-ID: adf8gasdd7fld@discrete.sampling.org
     CSeq: 1 INVITE
     Retry-After: Sun, 9 Jul 2000 11:59:00 GMT
     Content-Length: 0
```

M12 ACK sip:schockley@transistor.com SIP/2.0
```
     Via: SIP/2.0/UDP discrete.sampling.org:5060;branch=z9hG4bK5654
     Max-Forwards: 70
     From: Shannon <sip:shannon@sampling.org>;tag=cgdf4
     To: Schockley <sip:shockley@transistor.com>;tag=1
     Call-ID: adf8gasdd7fld@discrete.sampling.org
     CSeq: 1 ACK
     Content-Length: 0
```

M13 ACK sip:schockley@transistor.com SIP/2.0
```
     Via: SIP/2.0/UDP 9.8.7.6:5060;branch=z9hG4bK5.1
     Via: SIP/2.0/UDP discrete.sampling.org:5060;branch=z9hG4bK5654
     Max-Forwards: 69
     From: Shannon <sip:shannon@sampling.org>;tag=cgdf4
     To: Schockley <sip:shockley@transistor.com>;tag=1
     Call-ID: adf8gasdd7fld@discrete.sampling.org
```

```
CSeq: 1 ACK
Content-Length: 0
```

10.3   SIP to PSTN Call Through Gateway

In the example shown in Figure 10.3, the calling SIP phone places a telephone call to the PSTN through a PSTN gateway. The SIP phone collects the dialed digits and puts them into a SIP URI used in the Request-URI and the To header. The caller may have dialed either the globalized phone number 1-202-555-1313 or they may have just dialed a local number 555-1313, and the SIP phone added the assumed country code and area code to produce the globalized URI. The SIP phone has been preconfigured with the IP address of

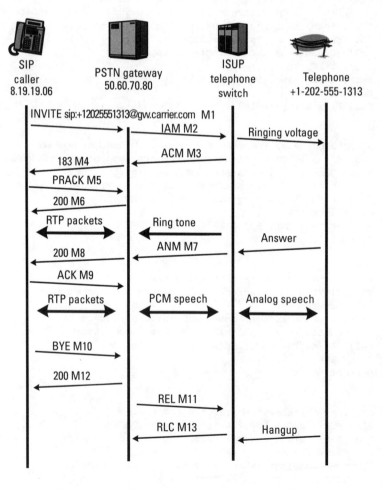

Figure 10.3 SIP to PSTN call through gateway.

the PSTN gateway, so it is able to send the `INVITE` directly to `gw.carrier.com`. The gateway initiates the call into the PSTN by selecting an SS7 ISUP trunk to the next telephone switch in the PSTN. The dialed digits from the `INVITE` are mapped into the ISUP IAM. The ISUP Address Complete Message (ACM) is sent back by the PSTN to indicate that the trunk has been seized. Progress tones are generated in the one-way audio path established in the PSTN. In this example, ring tone is generated by the far end telephone switch. The gateway maps the ACM to the `183 Session Progress` response containing SDP indicating the RTP port that the gateway will bridge the audio from the PSTN. Upon reception of the 183, the caller's UAC begins receiving the RTP packets sent from the gateway and presents the audio to the caller so they know that the call is progressing in the PSTN.

The call completes when the called party answers the telephone, which causes the telephone switch to send an Answer Message (ANM) to the gateway. The gateway then cuts the PSTN audio connection through in both directions and sends a `200 OK` response to the caller. Because the RTP media path is already established, the gateway echoes the SDP in the `183` but causes no changes to the RTP connection. The UAC sends an ACK to complete the SIP signaling exchange. Because there is no equivalent message in ISUP, the gateway absorbs the ACK.

The call terminates when the caller sends the `BYE` to the gateway. The gateway maps the `BYE` to the ISUP Release message or REL. The gateway sends the `200 OK` to the `BYE` and receives a RLC from the PSTN. These two messages have no dependency on each other; if, for some reason, either the SIP or PSTN network does not respond properly, one does not want resources held in the other network as a result.

```
M1    INVITE sip:+12025551313@gw.carrier.com;user=phone SIP/2.0
      Via: SIP/2.0/UDP 8.19.19.06:5060;branch=z9hG4bK4545
      Max-Forwards: 70
      From: <sip:filo.farnsworth@television.tv>;tag=12
      To: <sip:+12025551313@gw.carrier.com;user=phone>
      Call-ID: 49235243082018498@television.tv
      CSeq: 1 INVITE
      Supported: 100rel
      Contact: sip:filo.farnsworth@studio.television.tv
      Content-Type: application/sdp
      Content-Length: 154

      v=0
      o=FF 2890844535 2890844535 IN IP4 8.19.19.06
      s=-
      t=0 0
      c=IN IP4 8.19.19.06
      m=audio 5004 RTP/AVP 0 8               ⇐Two alternative codecs,
      a=rtpmap:0 PCMU/8000                        PCM μ-Law or
      a=rtpmap:8 PCMA/8000                          PCM A-Law
```

M2   IAM
     CdPN=202-555-1313, NPI=E.164,
     NOA=National   ⇐Gateway maps telephone
                                                          into called party number

M3   ACM

M4   SIP/2.0 183 Session Progress
     Via: SIP/2.0/UDP 8.19.19.06:5060;branch=z9hG4bK4545
     From: <sip:filo.farnsworth@television.tv>;tag=12
     To: <+12025551313@gw.carrier.com;user=phone>;tag=37 ⇐Tag and brackets
     Call-ID: 49235243082018498@television.tv
     CSeq: 1 INVITE
     RSeq: 08071
     Contact: <sip:50.60.70.80>
     Content-Type: application/sdp
     Content-Length: 139

     v=0
     o=Port1723 2890844535 2890844535 IN IP4 50.60.70.80
     s=-
     t=0 0
     c=IN IP4 50.60.70.80
     m=audio 62002 RTP/AVP 0   ⇐Gateway selects μ-Law codec
     a=rtpmap:0 PCMU/8000

M5   PRACK sip:50.60.70.80 SIP/2.0
     Via: SIP/2.0/UDP 8.19.19.06:5060;branch=z9hG4bK454
     Max-Forwards: 70
     From: <sip:filo.farnsworth@television.tv>;tag=37
     To: <sip:+12025551313@gw.carrier.com;user=phone>;tag=12
     Call-ID: 49235243082018498@television.tv
     CSeq: 2 PRACK
     Contact: sip:filo.farnsworth@studio.television.tv
     RAck: 08071 1 INVITE
     Content-Length: 0

M6   SIP/2.0 200 OK
     Via: SIP/2.0/UDP 8.19.19.06:5060;branch=z9hG4bK454
     From: <sip:filo.farnsworth@television.tv>;tag=37
     To: <sip:+12025551313@gw.carrier.com;user=phone>;tag=12
     Call-ID: 49235243082018498@television.tv
     CSeq: 2 PRACK

M7   ANM

```
M8    SIP/2.0 200 OK
      Via: SIP/2.0/UDP 8.19.19.06:5060;branch=z9hG4bK4545
      From: <sip:filo.farnsworth@television.tv>;tag=12
      To: <+12025551313@gw.carrier.com;user=phone>;tag=37
      Call-ID: 49235243082018498@television.tv
      CSeq: 1 INVITE
      Contact: <sip:50.60.70.80>
      Content-Type: application/sdp
      Content-Length: 139

      v=0
      o=Port1723 2890844535 2890844535 IN IP4 50.60.70.80
      s=-
      t=0 0
      c=IN IP4 50.60.70.80
      m=audio 62002 RTP/AVP 0
      a=rtpmap:0 PCMU/8000


M9    ACK sip:50.60.70.80 SIP/2.0
      Via: SIP/2.0/UDP 8.19.19.06:5060;branch=z9hG4bKfgrw
      Max-Forwards: 70
      From: <sip:filo.farnsworth@television.tv>;tag=12
      To: <+12025551313@gw.carrier.com;user=phone>;tag=37
      Call-ID: 49235243082018498@television.tv
      CSeq: 1 ACK



M10   BYE sip:50.60.70.80 SIP/2.0
      Via: SIP/2.0/UDP 8.19.19.06:5060;branch=z9hG4bK321
      Max-Forwards: 70
      From: <sip:filo.farnsworth@television.tv>;tag=12
      To: <+12025551313@gw.carrier.com;user=phone>;tag=37
      Call-ID: 49235243082018498@television.tv
      CSeq: 3 BYE                                  ⇐CSeq incremented


M11   REL
      CauseCode=16 Normal Clearing


M12   SIP/2.0 200 OK
      Via: SIP/2.0/UDP 8.19.19.06:5060;branch=z9hG4bK321
      From: <sip:filo.farnsworth@television.tv>;tag=12
      To: <+12025551313@gw.carrier.com;user=phone>;tag=37
      Call-ID: 49235243082018498@television.tv
      CSeq: 3 BYE


M13  RLC
```

10.4   PSTN to SIP Call Through Gateway

Figure 10.4 shows a call originating from a telephone in the PSTN that terminates on a SIP phone in the Internet. The compact form of SIP is used throughout the example. Note that there is no compact form for CSeq or
Max-Forwards.

```
M1    Setup
      CdPN=6512345, NPI=E.164,
      NOA=International              ⇐Dialed telephone number
      CgPN=4567890, NPI=E.164,
      NOA=International              ⇐PSTN caller's number


M2    INVITE sip:+6512345@incoming.com;user=phone SIP/2.0
      v: SIP/2.0/UDP 65.3.4.1:5060;branch=z9hG4bK343  ⇐Compact form of headers
      Max-Forwards: 70
      f: <sip:+45.67890@incoming.com;user=phone>;tag=6a589b1   ⇐includes tag
      t: <sip:+65.12345@incoming.com;user=phone>
      i: a3-65-99-1d@65.3.4.1
      CSeq: 1 INVITE
      m: sip:gw3.incoming.com
      c: application/sdp
      l: 126

      v=0
```

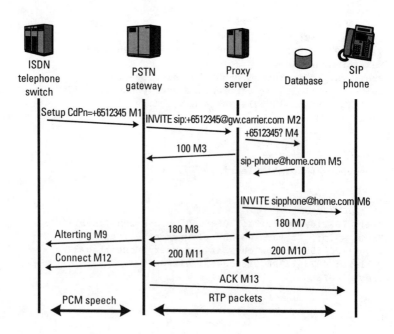

Figure 10.4 PSTN to SIP phone through gateway.

```
o=- 2890844535 2890844535 IN IP4 65.3.4.1
s=-
t=0 0
c=IN IP4 65.3.4.1
m=audio 62432 RTP/AVP 0
a=rtpmap:0 PCMU/8000
```

M3   SIP/2.0 100 Trying
```
v: SIP/2.0/UDP 65.3.4.1:5060;branch=z9hG4bK343
f: <sip:+45.67890@incoming.com;user=phone>;tag=6a589b1
t: sip:+65.12345@incoming.com;user=phone
i: a3-65-99-1d@65.3.4.1
CSeq: 1 INVITE
```

M4   Service Query: +65-12345

M5   Location Service Response:
     sip:user@home.com   ⇐Number maps to SIP URI

M6   INVITE sip:user@home.com SIP/2.0
```
v: SIP/2.0/UDP 176.5.8.2:5060;branch=z9hG4bK942834822.1
v: SIP/2.0/UDP 65.3.4.1:5060;branch=z9hG4bK343
Max-Forwards: 69
f: <sip:+45.67890@incoming.com;user=phone>;tag=6a589b1
t: sip:+65.12345@incoming.com;user=phone
i: a3-65-99-1d@65.3.4.1
CSeq: 1 INVITE
m: sip:gw3.incoming.com
c: application/sdp
l: 126

v=0
o=- 2890844535 2890844535 IN IP4 65.3.4.1
s=-
t=0 0
c=IN IP4 65.3.4.1
m=audio 62432 RTP/AVP 0
a=rtpmap:0 PCMU/8000
```

M7   SIP/2.0 180 Ringing
```
v: SIP/2.0/UDP 176.5.8.2:5060;branch=z9hG4bK942834822.1
v: SIP/2.0/UDP 65.3.4.1:5060;branch=z9hG4bK343
f: <sip:+45.67890@incoming.com;user=phone>;tag=6a589b1
t: <sip:+65.12345@incoming.com;user=phone>;tag=8657
i: a3-65-99-1d@65.3.4.1
m: sip:user@client.home.com
CSeq: 1 INVITE
```

```
M8    SIP/2.0 180 Ringing
      v: SIP/2.0/UDP 65.3.4.1:5060;branch=z9hG4bK343
      f: <sip:+45.67890@incoming.com;user=phone>;tag=6a589b1
      t: <sip:+65.12345@incoming.com;user=phone>;tag=8657
      i: a3-65-99-1d@65.3.4.1
      CSeq: 1 INVITE


M9    Alerting


M10   SIP/2.0 200 OK
      v: SIP/2.0/UDP 176.5.8.2:5060;branch=z9hG4bK942834822.1
      v: SIP/2.0/UDP 65.3.4.1:5060;branch=z9hG4bK343
      f: <sip:+45.67890@incoming.com;user=phone>;tag=6a589b1
      t: <sip:+65.12345@incoming.com;user=phone>;tag=8657
      i: a3-65-99-1d@65.3.4.1
      CSeq: 1 INVITE
      m: sip:user@client.home.com
      c: application/sdp
      l: 125

      v=0
      o=- 2890844565 2890844565 IN IP4 7.8.9.10
      s=-
      t=0 0
      c=IN IP4 7.8.9.10
      m=audio 5004 RTP/AVP 0
      a=rtpmap:0 PCMU/8000


M11   SIP/2.0 200 OK
      v: SIP/2.0/UDP 65.3.4.1:5060;branch=z9hG4bK343
      f: <sip:+45.67890@incoming.com;user=phone>;tag=6a589b1
      t: <sip:+65.12345@incoming.com;user=phone>;tag=8657
      i: a3-65-99-1d@65.3.4.1
      CSeq: 1 INVITE
      M: sip:user@home.com
      c: application/sdp
      l: 125

      v=0
      o=- 2890844565 2890844565 IN IP4 7.8.9.10
      s=-
      t=0 0
      c=IN IP4 7.8.9.10
      m=audio 5004 RTP/AVP 0
      a=rtpmap:0 PCMU/8000


M12   Connect


M13   ACK sip:user@home.com SIP/2.0
```

```
v: SIP/2.0/UDP 65.3.4.1:5060;branch=z9hG4bK453
Max-Forwards: 70
f: <sip+45.67890@incoming.com;user=phone>;tag=6a589b1
t: <sip+65.12345@incoming.com;user=phone>;tag=8657
i: a3-65-99-1d@65.3.4.1
CSeq: 1 ACK
```

10.5   Parallel Search

In this example the caller receives multiple possible locations for the called party from a redirect server. Instead of trying the locations one at a time, the user agent implements a parallel search for the called party by simultaneously sending the INVITE to three different locations, as shown in Figure 10.5. The SIP specification gives an example of this behavior in a proxy server, which is called a forking proxy.

In this example the first location responds with a 404 Not Found response. The second location responds with a 180 Ringing response, while the third location returns a 180 Ringing then a 200 OK response. The caller then sends an ACK to the third location to establish the call. Because one successful response has been received, a CANCEL is sent to the second location to terminate the search. The second location sends a 200 OK to the CANCEL and a 487 Request Terminated to the INVITE. This example shows some customized reason phrases in messages M7, M10, and M11.

```
M1    INVITE sip:faraday@effect.org;user=ip SIP/2.0
      Via: SIP/2.0/UDP 7.9.18.12:60000;branch=z9hG4bK3     ⇐Port 60000 is used
      Max-Forwards: 70
      From: J.C. Maxwell <sip:james.maxwell@kings.cambridge.edu.uk>;
       tag=4
      To: <sip:faraday@effect.org;user=ip>
      Call-ID: mNjdwWjkBfWrd@7.9.18.12
      CSeq: 54 INVITE                    ⇐CSeq initialized to 54
      Contact:<sip:james.maxwell@kings.cambridge.edu.uk>
      Content-Type: application/sdp
      Content-Length: 129

      v=0
      o=max 2890844521 2890844521 IN IP4 7.9.18.12
      s=-
      t=0 0
      c=IN IP4 7.9.18.12
      m=audio 32166 RTP/AVP 4
      a=rtpmap:4 DVI/8000


M2    SIP/2.0 300 Multiple locations                  ⇐Redirect server returns
      Via: SIP/2.0/UDP 7.9.18.12:60000;branch=z9hG4bK3        three locations
```

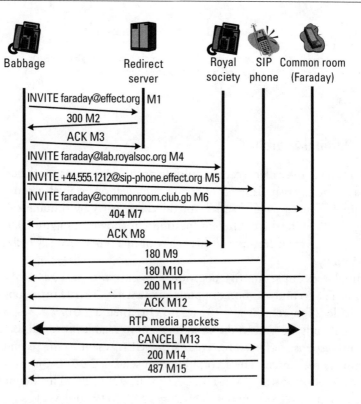

Figure 10.5 Parallel search example.

```
         From: J.C. Maxwell <sip:james.maxwell@kings.cambridge.edu.uk>
          ;tag=4
         To:<sip:faraday@effect.org;user=ip>;tag=1024
         Call-ID: mNjdwWjkBfWrd@7.9.18.12
         CSeq: 54 INVITE
         Contact:<sip:faraday@lab.royalsoc.gb>
         Contact:<sip:+44.555.1212@sip-phone.effect.org;user=phone>
         Contact: <sip:michael.faraday@commonroom.club.gb>


    M3   ACK sip:faraday@effect.org;user=ip
         Via: SIP/2.0/UDP 7.9.18.12:60000;branch=z9hG4bK3
         Max-Forwards: 70
         From: J.C. Maxwell <sip:james.maxwell@kings.cambridge.edu.uk>
          ;tag=4
         To: <sip:faraday@effect.org;user=ip>;tag=1024
         Call-ID: mNjdwWjkBfWrd@7.9.18.12
         CSeq: 54 INVITE


    M4   INVITE sip:faraday@lab.royalsoc.gb SIP/2.0
         Via: SIP/2.0/UDP 7.9.18.12:60000;branch=z9hG4bK1
```

```
Max-Forwards: 70
From: J.C. Maxwell <sip:james.maxwell@kings.cambridge.edu.uk>
 ;tag=5
To: <sip:faraday@effect.org;user=ip>            ⇐Tag is not copied
Call-ID: mNjdwWjkBfWrd@7.9.18.12                 ⇐Call-ID unchanged
CSeq: 55 INVITE                                  ⇐CSeq incremented
Contact: <sip:james.maxwell@kings.cambridge.edu.uk>
Content-Type: application/sdp
Content-Length: 129

v=0
o=max 2890844521 2890844521 IN IP4 7.9.18.12
s=-
t=0 0
c=IN IP4 7.9.18.12
m=audio 32166 RTP/AVP 4
a=rtpmap:4 DVI/8000
```

M5   INVITE sip:+44.555.1212@sip-phone.effect.org;user=phone SIP/2.0
```
Via: SIP/2.0/UDP 7.9.18.12:60000;branch=z9hG4bK2
Max-Forwards: 70
From: J.C. Maxwell <sip:james.maxwell@kings.cambridge.edu.uk>
 ;tag=5
To: <sip:faraday@effect.org;user=ip>
Call-ID: mNjdwWjkBfWrd@7.9.18.12
CSeq: 55 INVITE
Contact: <sip:james.maxwell@kings.cambridge.edu.uk>
Content-Type: application/sdp
Content-Length: 129

v=0
o=max 2890844521 2890844521 IN IP4 7.9.18.12
s=-
t=0 0
c=IN IP4 7.9.18.12
m=audio 32166 RTP/AVP 4
a=rtpmap:4 DVI/8000
```

M6   INVITE sip:faraday@commonroom.club.gb SIP/2.0
```
Via: SIP/2.0/UDP 7.9.18.12:60000;branch=z9hG4bK3
Max-Forwards: 70
From: J.C. Maxwell <sip:james.maxwell@kings.cambridge.edu.uk>
 ;tag=5
To: <sip:faraday@effect.org;user=ip>
Call-ID: mNjdwWjkBfWrd@7.9.18.12
CSeq: 55 INVITE
Contact: <sip:james.maxwell@kings.cambridge.edu.uk>
Content-Type: application/sdp
Content-Length: 129

v=0
o=max 2890844521 2890844521 IN IP4 7.9.18.12
```

```
       s=-
       t=0 0
       c=IN IP4 7.9.18.12
       m=audio 32166 RTP/AVP 4
       a=rtpmap:4 DVI/8000


M7     SIP/2.0 404 The member you have requested is not available
       Via: SIP/2.0/UDP 7.9.18.12:60000;branch=z9hG4bK1
       From: J.C. Maxwell <sip:james.maxwell@kings.cambridge.edu.uk>
        ;tag=5
       To: <sip:faraday@effect.org;user=ip>;tag=f6
       Call-ID: mNjdwWjkBfWrd@7.9.18.12
       CSeq: 55 INVITE


M8     ACK sip:faraday@lab.royalsoc.gb SIP/2.0
       Via: SIP/2.0/UDP 7.9.18.12:60000z9hG4bK1
       Max-Forwards: 70
       From: J.C. Maxwell <sip:james.maxwell@kings.cambridge.edu.uk>
        ;tag=5
       To: <sip:faraday@effect.org;user=ip>;tag=f6
       Call-ID: mNjdwWjkBfWrd@7.9.18.12
       CSeq: 55 ACK


M9     SIP/2.0 180 Ringing
       Via: SIP/2.0/UDP 7.9.18.12:60000;branch=z9hG4bK2
       From: J.C. Maxwell <sip:james.maxwell@kings.cambridge.edu.uk>
        ;tag=5
       To: <sip:faraday@effect.org;user=ip>;tag=6321
       Call-ID: mNjdwWjkBfWrd@7.9.18.12
       Contact: <sip:+44.555.1212@sip-phone.effect.org>
       CSeq: 55 INVITE


M10    SIP/2.0 180 Please wait while we locate Mr. Faraday
       Via: SIP/2.0/UDP 7.9.18.12:60000;branch=z9hG4bK3
       From: J.C. Maxwell <sip:james.maxwell@kings.cambridge.edu.uk>
        ;tag=5
       To: <sip:faraday@effect.org;user=ip>;tag=531
       Call-ID: mNjdwWjkBfWrd@7.9.18.12
       Contact: <sip:faraday@commonroom.club.gb>
       CSeq: 55 INVITE


M11    SIP/2.0 200 Mr. Faraday at your service?
       Via: SIP/2.0/UDP 7.9.18.12:60000;branch=z9hG4bK3
       From: J.C. Maxwell <sip:james.maxwell@kings.cambridge.edu.uk>
        ;tag=5
       To: <sip:faraday@effect.org;user=ip>;tag=531
       Call-ID: mNjdwWjkBfWrd@7.9.18.12
       CSeq: 55 INVITE
       User-Agent: PDV v4
```

```
       Contact: <sip:faraday@commonroom.club.gb>
       Content-Type: application/sdp
       Content-Length: 131

       v=0
       o=max 2890844521 2890844521 IN IP4 6.22.17.89
       t=0 0
       c=IN IP4 6.22.17.89
       m=audio 43782 RTP/AVP 4
       a=rtpmap:4 DVI/8000
```

```
M12    ACK sip:faraday@commonroom.club.gb;user=ip SIP/2.0
       Via: SIP/2.0/UDP 7.9.18.12:60000;branch=z9hG4bK3
       Max-Forwards: 70
       From: J.C. Maxwell <sip:james.maxwell@kings.cambridge.edu.uk>
        ;tag=5
       To: <sip:faraday@effect.org;user=ip>;tag=531
       Call-ID: mNjdwWjkBfWrd@7.9.18.12
       CSeq: 55 ACK
```

```
M13    CANCEL sip:+44.555.1212@sip-phone.effect.org;user=phone SIP/2.0
       Via: SIP/2.0/UDP 7.9.18.12:60000;branch=z9hG4bK2          ⇐Cancels search
       Max-Forwards: 70
       From: J.C. Maxwell <sip:james.maxwell@kings.cambridge.edu.uk>
        ;tag=5
       To: <sip:faraday@effect.org;user=ip>;tag=6321
       Call-ID: mNjdwWjkBfWrd@7.9.18.12
       CSeq: 55 CANCEL                               ⇐CSeq not incremented
                                                      Method set to CANCEL
```

```
M14    SIP/2.0 200 OK                               ⇐CANCEL acknowledged
       Via: SIP/2.0/UDP 7.9.18.12:60000;branch=z9hG4bK2
       From: J.C. Maxwell <sip:james.maxwell@kings.cambridge.edu.uk>
        ;tag=5
       To: <sip:faraday@effect.org;user=ip>;tag=6321
       Call-ID: mNjdwWjkBfWrd@7.9.18.12
       CSeq: 55 CANCEL
```

```
M15    SIP/2.0 487 Request Terminated               ⇐Final response to INVITE
       Via: SIP/2.0/UDP 7.9.18.12:60000;branch=z9hG4bK2
       From: J.C. Maxwell <sip:james.maxwell@kings.cambridge.edu.uk>
        ;tag=5
       To: <sip:faraday@effect.org;user=ip>;tag=6321
       Call-ID: mNjdwWjkBfWrd@7.9.18.12
       CSeq: 55 INVITE
```

```
M16    ACK SIP/2.0
       Via: SIP/2.0/UDP 7.9.18.12:60000;branch=z9hG4bK2
       From: J.C. Maxwell <sip:james.maxwell@kings.cambridge.edu.uk>
        ;tag=5
```

```
To: <sip:faraday@effect.org;user=ip>;tag=6321
Call-ID: mNjdwWjkBfWrd@7.9.18.12
CSeq: 55 ACK
```

10.6   H.323 to SIP Call

In this example, a H.323 terminal calls a SIP-enabled PC through a H.323/SIP gateway. The gateway does signaling translation between the protocols but allows the two end points to exchange media packets directly with each other. The full details of SIP/H.323 interworking are being developed in the SIP working group [4].

In this example, shown in Figure 10.6, the initial message exchange is between the calling H.323 terminal and the H.323 gatekeeper. The gatekeeper resolves the H.323 alias into an address served by the H.323/SIP gateway. The ACF response indicates that gatekeeper-routed signaling is required, so the Q.931 and H.245 TCP connections are opened to the gatekeeper, which opens TCP connections to the gateway. The calling H.323 terminal sends a Q.931 Setup message to the gatekeeper, which proxies it to the H.323/SIP gateway. The gateway then looks up the H.323 alias and resolves it to the SIP URI of the called party. It constructs an INVITE from the Setup message and forwards it to a SIP proxy, which forwards it to the called party. Note that because the Setup message does not contain any media information, the INVITE does not contain any media information either. The called party sends a 180 Ringing then a 200 OK to indicate that the call has been answered. The media information present in the SDP message body is stored by the gateway, which sends Alerting and Connect messages.

Messages are sent to the gatekeeper, which proxies them to the calling H.323 terminal. The gateway holds off sending the ACK response to the INVITE until the H.245 media exchange is completed between the H.323 terminal and the gateway. Once that is complete, the negotiated media capabilities are returned in the ACK and the media session begins.

```
M1    ARQ
      address(h323alias=Stibitz)


M2    ACF
      gatekeeper routed signaling


M3    Setup
      Cd address(h323alias=Stibitz)
      Cg address(h323alias=Burroughs)


M4    Setup
```

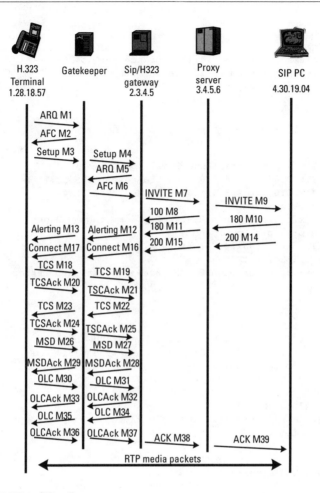

Figure 10.6 H.323 to SIP call.

```
Cd address(h323alias=Stibitz)
Cg address(h323alias=Burroughs)
```

M5   ARQ

M6   ACF

M7   INVITE sip:stibitz@proxy.com SIP/2.0
      Via: SIP/2.0/TCP 2.3.4.5:4344;branch=z9hG4bK544 ⇐TCP used for transport
      Max-Forwards: 70
      From: <sip:burroughs@h323-gateway.com>;tag=3q526g45
      To: <sip:stibitz@proxy.com>
      Call-ID: 526272332146783569054
      CSeq: 43252 INVITE   ⇐CSeq initialized to 43252
      Contact: <sip:burroughs@h323-gateway.com;transport=tcp>

```
        Content-Length: 0
M8      100 Trying
        Via: SIP/2.0/TCP 2.3.4.5:4344;branch=z9hG4bK544
        From: <sip:burroughs@h323-gateway.com>
        To: <sip:stibitz@proxy.com>
        Call-ID: 526272332146783569054
        CSeq: 43252 INVITE
        Content-Length: 0


M9      INVITE sip:gstibitz123@snailmail.com SIP/2.0
        Via: SIP/2.0/TCP 3.4.5.6:5060;branch=z9hG4bK452.1
        Via: SIP/2.0/TCP 2.3.4.5:4344;branch=z9hG4bK544
        Max-Forwards: 69
        From: <sip:burroughs@h323-gateway.com>;tag=3q526g45
        To: <sip:stibitz@proxy.com>
        Call-ID: 526272332146783569054
        CSeq: 43252 INVITE
        Contact: <sip:burroughs@h323-gateway.com;transport=tcp>


M10     SIP/2.0 180 Ringing
        Via: SIP/2.0/TCP 3.4.5.6:5060; branch=z9hG4bK452.1
        Via: SIP/2.0/TCP 2.3.4.5:4344;branch=z9hG4bK544
        From: <sip:burroughs@h323-gateway.com>;tag=3q526g45
        To: <sip:stibitz@proxy.com>;tag=1926
        Call-ID: 526272332146783569054
        CSeq: 43252 INVITE
        Contact: <sip:gstibitz123@snailmail.com;transport=tcp>
        Content-Length: 0


M11     SIP/2.0 180 Ringing
        Via: SIP/2.0/TCP 2.3.4.5:4344;branch=z9hG4bK544
        From: <sip:burroughs@h323-gateway.com>
        To: <sip:stibitz@proxy.com>;tag=1926
        Call-ID: 526272332146783569054
        CSeq: 43252 INVITE
        Contact: <sip:gstibitz123@snailmail.com;transport=tcp>
        Content-Length: 0


M12     Alerting


M13     Alerting


M14     SIP/2.0 200 OK
        Via: SIP/2.0/TCP 3.4.5.6:5060; branch=z9hG4bK452.1
        Via: SIP/2.0/TCP 2.3.4.5:4344;branch=z9hG4bK544
        From: <sip:burroughs@h323-gateway.com>;tag=3q526g45
        To: <sip:stibitz@proxy.com>;tag=1926
```

```
        Call-ID: 526272332146783569054
        CSeq: 43252 INVITE
        Contact: <sip:gstibitz123@snailmail.com;transport=tcp>
        Content-Type: application/sdp
        Content-Length: 134

        v=0
        o=George 2890844576 2890844576 IN IP4 4.30.19.04
        s=-
        t=0 0
        c=IN IP4 4.30.19.04
        m=audio 5004 RTP/AVP 0
        a=rtpmap:0 PCMU/8000


M15   SIP/2.0 200 OK
        Via: SIP/2.0/TCP 2.3.4.5:4344;branch=z9hG4bK544
        From: <sip:burroughs@h323-gateway.com>;tag=3q526g45
        To: <sip:stibitz@proxy.com>;tag=1926
        Call-ID: 526272332146783569054
        CSeq: 43252 INVITE
        Contact: <sip:gstibitz123@snailmail.com>
        Content-Type: application/sdp
        Content-Length: 134

        v=0
        o=George 2890844576 2890844576 IN IP4 4.30.19.04
        s=-
        t=0 0
        c=IN IP4 4.30.19.04
        m=audio 5004 RTP/AVP 0
        a=rtpmap:0 PCMU/8000


M16   Connect


M17   Connect


M18   TerminalCapabilitySet


M19   TerminalCapabilitySet


M20   TerminalCapabilitySetAck


M21   TerminalCapabilitySetAck


M22   TerminalCapabilitySet
```

```
M23  TerminalCapabilitySet


M24  TerminalCapabilitySetAck


M25  TerminalCapabilitySetAck


M26  MasterSlaveDetermination


M27  MasterSlaveDetermination


M28  MasterSlaveDeterminationAck


M29  MasterSlaveDeterminationAck


M30  OpenLogicalChannel
     g711uLaw 1.28.18.57 60002


M31  OpenLogicalChannel
     g711uLaw 1.28.18.57 60002


M32  OpenLogicalChannelAck


M33  OpenLogicalChannelAck
     Call Flow Examples 181


M34  OpenLogicalChannel
     g711uLaw 4.30.19.04 5004


M35  OpenLogicalChannel
     g711uLaw 4.30.19.04 5004


M36  OpenLogicalChannelAck


M37  OpenLogicalChannelAck


M38  ACK sip:gstibitz123@snailmail.com SIP/2.0
     Via: SIP/2.0/TCP 2.3.4.5:4344;branch=z9hG4bK32
     Max-Forwards: 70
```

```
From: <sip:burroughs@h323-gateway.com>;tag=3q526g45
To: <sip:stibitz@proxy.com>;tag=1926
Call-ID: 526272332146783569054
CSeq: 43252 ACK
Content-Type: application/sdp
Content-Length: 130

v=0
o=- 2890844577 2890844577 IN IP4 1.28.18.57
s=-
t=0 0
c=IN IP4 1.28.18.57
m=audio 60002 RTP/AVP 0
a=rtpmap:0 PCMU/8000
```

```
M39  ACK sip:gstibitz123@snailmail.com SIP/2.0
     Via: SIP/2.0/TCP 3.4.5.6:5060;branch=z9hG4bK452.1
     Via: SIP/2.0/TCP 2.3.4.5:4344;branch=z9hG4bK532
     Max-Forwards: 69
     From: <sip:burroughs@h323-gateway.com>;tag=3q526g45
     To: <sip:stibitz@proxy.com>;tag=1926
     Call-ID: 526272332146783569054
     CSeq: 43252 ACK
     Content-Type: application/sdp
     Content-Length: 130

     v=0
     o=- 2890844577 2890844577 IN IP4 1.28.18.57
     s=-
     t=0 0
     c=IN IP4 1.28.18.57
     m=audio 60002 RTP/AVP 0
     a=rtpmap:0 PCMU/8000
```

10.7 3GPP Wireless Call Flow

An example SIP 3GPP call flow is shown in Figure 10.7. Some of the 3GPP P-headers are not shown with their actual values—instead, "..." is listed. Also, some call flows show SDP in the PRACK, but this is only needed if another SDP offer is made—in this case, the initial offer is accepted. Also, PRACKs are not shown in response to the 180 Ringing response since it does not carry any SDP and does not need to be delivered reliably. In this call flow, the use of QoS preconditions is shown (as indicated by the Require: precondition header field and the QoS attributes in the SDP). After the initial offer and answer in the INVITE and 183, the QoS setup is performed. When it has been done successfully, each side indicates in the UPDATE and 200 OK response in the SDP that the setup is complete. Only after that does the called UE begin alerting and sends a 180 Ringing response.

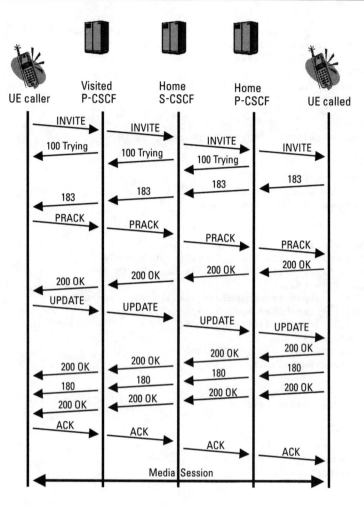

Figure 10.7 3GPP SIP call flow.

```
INVITE tel:+1-314-555-1234 SIP/2.0
Via: SIP/2.0/UDP [::0000:1111:2222:3333]:2343;comp=sigcomp
 ;branch=z9hG4bKmp17a43
Max-Forwards: 70
Route: <sip:pcscf.visitednet.com:4323;lr>,
<sip:scscf.homenet.com;lr>
To: <tel:+1-314-555-1234>
From: <sip:mobile1@homenet.com>;tag=34199357834
P-Preferred-Identity: "Alice" <sip:mobile1@homnet.com>
P-Access-Network-Info: 3GPP-UTRAN-TDD; utran-cell-id-
3gpp=544542332
Privacy: none
Call-ID: 7o5452gfvtyrfghdgrtt6900gh
CSeq: 15 INVITE
```

```
Contact: <sip:[::0000:1111:2222:3333]:2343;comp=sigcomp>
Supported: 100rel
Require: precondition, sec-agree
Proxy-Require: sec-agree
Allow: INVITE, ACK, CANCEL, BYE, PRACK, MESSSAGE, REFER, UPDATE
Security-Verify: ipsec-3gpp; q=0.1; alg=hmac-sha-1-96; spi=34341
  ;port1=4323
Content-Type: application/sdp
Content-Length: 159

v=0
o=- 2890844526 2890844526 IN IP4 [::0000:1111:2222:3333]
s=Phone Call
t=0 0
c=IN IP6 [::0000:1111:2222:3333]
m=audio 49170 RTP/AVP 97 96
b=AS:25.4
a=curr:qos local none
a=curr:qos remote none
a=des:qos mandatory local sendrecv
a=des:qos none remote sendrecv
a=rtpmap:97 AMR
a=fmtp:97 mode-set=0,2,5,7 ;maxframes=2
a=rtpmap:96 telephone-event


SIP/2.0 100 Trying
Via: SIP/2.0/UDP
  [::0000:1111:2222:3333]:2343;branch=z9hG4bKmp17a43
To: <tel:+1-314-555-1234>
From: <sip:mobile1@homenet.com>;tag=34199357834
Call-ID: 7o5452gfvtyrfghdgrtt6900gh
CSeq: 15 INVITE
Content-Length: 0


INVITE tel:+1-314-555-1234 SIP/2.0
Via: SIP/2.0/UDP pcscf.visitednet.com;branch=z9hG4bK7866
Via: SIP/2.0/UDP [::0000:1111:2222:3333]:2343;comp=sigcomp
  ;branch=z9hG4bKmp17a43
Max-Forwards: 69
Route: <sip:scscf.homenet.com;lr>
Record-Route: <sip:pcscf.visitednet.com;lr>
To: <tel:+1-314-555-1234>
From: <sip:mobile1@homenet.com>;tag=34199357834
P-Asserted-Identity: "Alice" <sip:mobile1@homnet.com>
P-Access-Network-Info: 3GPP-UTRAN-TDD;
utran-cell-id-3gpp=544542332
Privacy: none
P-Charging-Vector: icid-value=914052945
  ;icid-generated-at=[::01:34:3F:34]
Call-ID: 7o5452gfvtyrfghdgrtt6900gh
CSeq: 15 INVITE
Contact: <sip:[::0000:1111:2222:3333]:2343;comp=sigcomp>
```

```
Supported: 100rel
Require: precondition
Allow: INVITE, ACK, CANCEL, BYE, PRACK, MESSSAGE, REFER, UPDATE
Security-Verify: ipsec-3gpp; q=0.1; alg=hmac-sha-1-96; spi=34341
 ;port1=4323
Content-Type: application/sdp
Content-Length: 159

v=0
o=- 2890844526 2890844526 IN IP4 [::0000:1111:2222:3333]
s=Phone Call
t=0 0
c=IN IP6 [::0000:1111:2222:3333]
m=audio 49170 RTP/AVP 97 96
b=AS:25.4
a=curr:qos local none
a=curr:qos remote none
a=des:qos mandatory local sendrecv
a=des:qos none remote sendrecv
a=rtpmap:97 AMR
a=fmtp:97 mode-set=0,2,5,7 ;maxframes=2
a=rtpmap:96 telephone-event


SIP/2.0 100 Trying
Via: SIP/2.0/UDP pcscf.visitednet.com;branch=z9hG4bK7866
Via: SIP/2.0/UDP [::0000:1111:2222:3333]:2343;comp=sigcomp
 ;branch=z9hG4bKmp17a43
To: <tel:+1-314-555-1234>
From: <sip:mobile1@homenet.com>;tag=34199357834
Call-ID: 7o5452gfvtyrfghdgrtt6900gh
CSeq: 15 INVITE
Content-Length: 0


INVITE sip:mobile2@homenet.com SIP/2.0
Via: SIP/2.0/UDP scscf.homenet.com;branch=z9hG4bK34fd2
Via: SIP/2.0/UDP pcscf.visitednet.com;branch=z9hG4bK7866
Via: SIP/2.0/UDP [::0000:1111:2222:3333]:2343;comp=sigcomp
 ;branch=z9hG4bKmp17a43
Max-Forwards: 68
Record-Route: <scscf.homenet.com>, <sip:pcscf.visitednet.com;lr>
To: <tel:+1-314-555-1234>
From: <sip:mobile1@homenet.com>;tag=34199357834
P-Asserted-Identity: "Alice" <sip:mobile1@homnet.com>,
 <tel:+1-972-555-4321>
P-Access-Network-Info: . . .
Privacy: none
P-Charging-Vector: . . .
Call-ID: 7o5452gfvtyrfghdgrtt6900gh
CSeq: 15 INVITE
Contact: <sip:[::0000:1111:2222:3333]:2343;comp=sigcomp>
Supported: 100rel
Require: precondition
```

```
Allow: INVITE, ACK, CANCEL, BYE, PRACK, MESSSAGE, REFER, UPDATE
Security-Verify: ipsec-3gpp; q=0.1; alg=hmac-sha-1-96; spi=34341
 ;port1=4323
Content-Type: application/sdp
Content-Length: 159

v=0
o=- 2890844526 2890844526 IN IP4 [::0000:1111:2222:3333]
s=Phone Call
t=0 0
c=IN IP6 [::0000:1111:2222:3333]
m=audio 49170 RTP/AVP 97 96
b=AS:25.4
a=curr:qos local none
a=curr:qos remote none
a=des:qos mandatory local sendrecv
a=des:qos none remote sendrecv
a=rtpmap:97 AMR
a=fmtp:97 mode-set=0,2,5,7 ;maxframes=2
a=rtpmap:96 telephone-event


SIP/2.0 100 Trying
Via: SIP/2.0/UDP scscf.homenet.com;branch=z9hG4bK34fd2
Via: SIP/2.0/UDP pcscf.visitednet.com;branch=z9hG4bK7866
Via: SIP/2.0/UDP [::0000:1111:2222:3333]:2343;comp=sigcomp
 ;branch=z9hG4bKmp17a43
To: <tel:+1-314-555-1234>
From: <sip:mobile1@homenet.com>;tag=34199357834
Call-ID: 7o5452gfvtyrfghdgrtt6900gh
CSeq: 15 INVITE
Content-Length: 0



INVITE sip:mobile2@homenet.com SIP/2.0
Via: SIP/2.0/UDP pcscf.homenet.com;branch=z9hG4bK42234
Via: SIP/2.0/UDP scscf.homenet.com;branch=z9hG4bK34fd2
Via: SIP/2.0/UDP pcscf.visitednet.com;branch=z9hG4bK7866
Via: SIP/2.0/UDP [::0000:1111:2222:3333]:2343;comp=sigcomp
 ;branch=z9hG4bKmp17a43
Max-Forwards: 67
Record-Route: <sip:pscsf.homenet.com;lr>, <scscf.homenet.com;lr>,
 <sip:pcscf.visitednet.com;lr>
To: <tel:+1-314-555-1234>
From: <sip:mobile1@homenet.com>;tag=34199357834
P-Asserted-Identity: "Alice" <sip:mobile1@homnet.com>,
 <tel:+1-972-555-4321>
P-Access-Network-Info: . . .
Privacy: none
P-Charging-Vector: . . .
P-Called-Party-ID: <sip:mobile2@homenet.com>
P-Media-Authorization: 4504958682459049502984084485948945
```

```
Call-ID: 7o5452gfvtyrfghdgrtt6900gh
CSeq: 15 INVITE
Contact: <sip:[::0000:1111:2222:3333]:2343;comp=sigcomp>
Supported: 100rel
Require: precondition
Allow: INVITE, ACK, CANCEL, BYE, PRACK, MESSSAGE, REFER, UPDATE
Security-Verify: ipsec-3gpp; q=0.1; alg=hmac-sha-1-96; spi=34341
 ;port1=4323
Content-Type: application/sdp
Content-Length: ...

v=0
o=- 2890844526 2890844526 IN IP4 [::0000:1111:2222:3333]
s=Phone Call
t=0 0
c=IN IP6 [::0000:1111:2222:3333]
m=audio 49170 RTP/AVP 97 96
b=AS:25.4
a=curr:qos local none
a=curr:qos remote none
a=des:qos mandatory local sendrecv
a=des:qos none remote sendrecv
a=rtpmap:97 AMR
a=fmtp:97 mode-set=0,2,5,7 ;maxframes=2
a=rtpmap:96 telephone-event


SIP/2.0 183 Session Progress
Via: SIP/2.0/UDP pcscf.homenet.com;branch=z9hG4bK42234
Via: SIP/2.0/UDP scscf.homenet.com;branch=z9hG4bK34fd2
Via: SIP/2.0/UDP pcscf.visitednet.com;branch=z9hG4bK7866
Via: SIP/2.0/UDP [::0000:1111:2222:3333]:2343;comp=sigcomp
 ;branch=z9hG4bKmp17a43
Record-Route: <sip:pscsf.homenet.com;lr>, <scscf.homenet.com;lr>,
 <sip:pcscf.visitednet.com;lr>
To: <tel:+1-314-555-1234>;tag=g41334
From: <sip:mobile1@homenet.com>;tag=34199357834
P-Access-Network-Info: . . .
Privacy: none
Call-ID: 7o5452gfvtyrfghdgrtt6900gh
CSeq: 15 INVITE
Contact: <sip:[::0000:aaaa:bbbb:cccc]:8767;comp=sigcomp>
Require: 100rel
Allow: INVITE, ACK, CANCEL, BYE, PRACK, MESSSAGE, REFER, UPDATE
RSeq: 98205
Content-Type: application/sdp
Content-Length: 159

v=0
o=- 28908445761 28908445761 IN IP6 [::0000:aaaa:bbbb:cccc]
s=-
t=0 0
c=IN IP6 [::0000:aaaa:bbbb:cccc]
m=audio 5434 RTP/AVP 97 96
```

```
b=AS:25.4
a=curr:qos local none
a=curr:qos remote none
a=des:qos mandatory local sendrecv
a=des:qos none remote sendrecv
a=rtpmap:97 AMR
a=fmtp:97 mode-set=0,2,5,7 ;maxframes=2
a=rtpmap:96 telephone-event


SIP/2.0 183 Session Progress
Via: SIP/2.0/UDP scscf.homenet.com;branch=z9hG4bK34fd2
Via: SIP/2.0/UDP pcscf.visitednet.com;branch=z9hG4bK7866
Via: SIP/2.0/UDP [::0000:1111:2222:3333]:2343;comp=sigcomp
 ;branch=z9hG4bKmp17a43
Record-Route: <sip:pscsf.homenet.com;lr>, <scscf.homenet.com;lr>,
 <sip:pcscf.visitednet.com;lr>
To: <tel:+1-314-555-1234>;tag=g41334
From: <sip:mobile1@homenet.com>;tag=34199357834
P-Access-Network-Info: 3GPP-UTRAN-TDD; utran-cell-id-3gpp=9459495
Privacy: none
P-Charging-Vector: . . .
P-Asserted-Identity: "Bob" <sip:mobile2@homenet.com>,
Call-ID: 7o5452gfvtyrfghdgrtt6900gh
CSeq: 15 INVITE
Contact: <sip:[::0000:aaaa:bbbb:cccc]:8767;comp=sigcomp>
Require: 100rel
Allow: INVITE, ACK, CANCEL, BYE, PRACK, MESSSAGE, REFER, UPDATE
RSeq: 98205
Content-Type: application/sdp
Content-Length: 159

v=0
o=- 28908445761 28908445761 IN IP6 [::0000:aaaa:bbbb:cccc]
s=-
t=0 0
c=IN IP6 [::0000:aaaa:bbbb:cccc]
m=audio 5434 RTP/AVP 97 96
b=AS:25.4
a=curr:qos local none
a=curr:qos remote none
a=des:qos mandatory local sendrecv
a=des:qos none remote sendrecv
a=rtpmap:97 AMR
a=fmtp:97 mode-set=0,2,5,7 ;maxframes=2
a=rtpmap:96 telephone-event


SIP/2.0 183 Session Progress
Via: SIP/2.0/UDP pcscf.visitednet.com;branch=z9hG4bK7866
Via: SIP/2.0/UDP [::0000:1111:2222:3333]:2343;comp=sigcomp
 ;branch=z9hG4bKmp17a43
Record-Route: <sip:pscsf.homenet.com;lr>, <scscf.homenet.com;lr>,
 <sip:pcscf.visitednet.com;lr>
```

```
To: <tel:+1-314-555-1234>;tag=g41334
From: <sip:mobile1@homenet.com>;tag=34199357834
P-Access-Network-Info: 3GPP-UTRAN-TDD; utran-cell-id-3gpp=9459495
Privacy: none
P-Charging-Vector: . . .
P-Asserted-Identity: "Bob" <sip:mobile2@homenet.com>,
Call-ID: 7o5452gfvtyrfghdgrtt6900gh
CSeq: 15 INVITE
Contact: <sip:[::0000:aaaa:bbbb:cccc]:8767;comp=sigcomp>
Require: 100rel
Allow: INVITE, ACK, CANCEL, BYE, PRACK, MESSSAGE, REFER, UPDATE
RSeq: 98205
Content-Type: application/sdp
Content-Length: 159

v=0
o=- 28908445761 28908445761 IN IP6 [::0000:aaaa:bbbb:cccc]
s=-
t=0 0
c=IN IP6 [::0000:aaaa:bbbb:cccc]
m=audio 5434 RTP/AVP 97 96
b=AS:25.4
a=curr:qos local none
a=curr:qos remote none
a=des:qos mandatory local sendrecv
a=des:qos none remote sendrecv
a=rtpmap:97 AMR
a=fmtp:97 mode-set=0,2,5,7 ;maxframes=2
a=rtpmap:96 telephone-event


SIP/2.0 183 Session Progress
Via: SIP/2.0/UDP [::0000:1111:2222:3333]:2343;comp=sigcomp
 ;branch=z9hG4bKmp17a43
Record-Route: <sip:pscsf.homenet.com;lr>, <scscf.homenet.com;lr>,
 <sip:pcscf.visitednet.com;lr>
To: <tel:+1-314-555-1234>;tag=g41334
From: <sip:mobile1@homenet.com>;tag=34199357834
P-Access-Network-Info: 3GPP-UTRAN-TDD; utran-cell-id-3gpp=9459495
Privacy: none
P-Charging-Vector: . . .
P-Asserted-Identity: "Bob" <sip:mobile2@homenet.com>,
P-Media-Authorization: 450988409440958209582020203058
Call-ID: 7o5452gfvtyrfghdgrtt6900gh
CSeq: 15 INVITE
Contact: <sip:[::0000:aaaa:bbbb:cccc]:8767;comp=sigcomp>
Require: 100rel
Allow: INVITE, ACK, CANCEL, BYE, PRACK, MESSSAGE, REFER, UPDATE
RSeq: 98205
Content-Type: application/sdp
Content-Length: 159

v=0
o=- 28908445761 28908445761 IN IP6 [::0000:aaaa:bbbb:cccc]
```

```
s=-
t=0 0
c=IN IP6 [::0000:aaaa:bbbb:cccc]
m=audio 5434 RTP/AVP 97 96
b=AS:25.4
a=curr:qos local none
a=curr:qos remote none
a=des:qos mandatory local sendrecv
a=des:qos none remote sendrecv
a=rtpmap:97 AMR
a=fmtp:97 mode-set=0,2,5,7 ;maxframes=2
a=rtpmap:96 telephone-event
```

```
PRACK sip:[::0000:aaaa:bbbb:cccc]:8767;comp=sigcomp SIP/2.0
Via: SIP/2.0/UDP [::0000:1111:2222:3333]:2343;comp=sigcomp
 ;branch=z9hG4bK454245
Max-Forwards: 70
Route: <sip:pcscf.visitednet.com;lr>, <scscf.homenet.com;lr>,
 <sip:pscsf.homenet.com;lr>
To: <tel:+1-314-555-1234>;tag=g41334
From: <sip:mobile1@homenet.com>;tag=34199357834
P-Access-Network-Info: . . .
Call-ID: 7o5452gfvtyrfghdgrtt6900gh
CSeq: 16 PRACK
Contact: <sip:[::0000:1111:2222:3333]:2343;comp=sigcomp>
Supported: 100rel
Require: precondition
Allow: INVITE, ACK, CANCEL, BYE, PRACK, MESSSAGE, REFER, UPDATE
RAck: 98205 15 INVITE
Content-Length: 0
```

```
PRACK sip:[::0000:aaaa:bbbb:cccc]:8767;comp=sigcomp SIP/2.0
Via: SIP/2.0/UDP pcscf.visitednet.com;branch=z9hG4bK565355
Via: SIP/2.0/UDP [::0000:1111:2222:3333]:2343;comp=sigcomp
 ;branch=z9hG4bK454245
Max-Forwards: 69
Route: <sip:pscsf.homenet.com;lr>, <sip:scscf.homenet.com;er>
To: <tel:+1-314-555-1234>;tag=g41334
From: <sip:mobile1@homenet.com>;tag=34199357834
P-Access-Network-Info: . . .
Call-ID: 7o5452gfvtyrfghdgrtt6900gh
CSeq: 16 PRACK
Contact: <sip:[::0000:1111:2222:3333]:2343;comp=sigcomp>
Supported: 100rel
Require: precondition
Allow: INVITE, ACK, CANCEL, BYE, PRACK, MESSSAGE, REFER, UPDATE
RAck: 98205 15 INVITE
Content-Length: 0
```

```
PRACK sip:[::0000:aaaa:bbbb:cccc]:8767;comp=sigcomp SIP/2.0
Via: SIP/2.0/UDP scscf.homenet.com;branch=z9hG4bKhj5fy
```

```
Via: SIP/2.0/UDP pcscf.visitednet.com;branch=z9hG4bK7866
Via: SIP/2.0/UDP [::0000:1111:2222:3333]:2343;comp=sigcomp
 ;branch=z9hG4bK454245
Max-Forwards: 68
To: <tel:+1-314-555-1234>;tag=g41334
From: <sip:mobile1@homenet.com>;tag=34199357834
Call-ID: 7o5452gfvtyrfghdgrtt6900gh
CSeq: 16 PRACK
Contact: <sip:[::0000:1111:2222:3333]:2343;comp=sigcomp>
Supported: 100rel
Require: precondition
Allow: INVITE, ACK, CANCEL, BYE, PRACK, MESSSAGE, REFER, UPDATE
RAck: 98205 15 INVITE <Route: <sip:pscsf.homenet.com;lr>
Content-Length: 0


PRACK sip:[::0000:aaaa:bbbb:cccc]:8767;comp=sigcomp SIP/2.0
Via: SIP/2.0/UDP pcscf.homenet.com;branch=z9hG4bK4524
Via: SIP/2.0/UDP scscf.homenet.com;branch=z9hG4bKhj5fy
Via: SIP/2.0/UDP pcscf.visitednet.com;branch=z9hG4bK7866
Via: SIP/2.0/UDP [::0000:1111:2222:3333]:2343;comp=sigcomp
 ;branch=z9hG4bK454245
Max-Forwards: 67
To: <tel:+1-314-555-1234>;tag=g41334
From: <sip:mobile1@homenet.com>;tag=34199357834
P-Access-Network-Info: 3GPP-UTRAN-TDD; utran-cell-id-
3gpp=544542332
Call-ID: 7o5452gfvtyrfghdgrtt6900gh
CSeq: 16 PRACK
Contact: <sip:[::0000:1111:2222:3333]:2343;comp=sigcomp>
Supported: 100rel
Require: precondition
Allow: INVITE, ACK, CANCEL, BYE, PRACK, MESSSAGE, REFER, UPDATE
RAck: 98205 15 INVITE
Content-Length: 0


SIP/2.0 200 OK
Via: SIP/2.0/UDP pcscf.homenet.com;branch=z9hG4bK4524
Via: SIP/2.0/UDP scscf.homenet.com;branch=z9hG4bKhj5fy
Via: SIP/2.0/UDP pcscf.visitednet.com;branch=z9hG4bK7866
Via: SIP/2.0/UDP [::0000:1111:2222:3333]:2343;comp=sigcomp
 ;branch=z9hG4bK454245
To: <tel:+1-314-555-1234>;tag=g41334
From: <sip:mobile1@homenet.com>;tag=34199357834
P-Access-Network-Info: 3GPP-UTRAN-TDD; utran-cell-id-
3gpp=544542332
Call-ID: 7o5452gfvtyrfghdgrtt6900gh
CSeq: 16 PRACK
Content-Length: 0


SIP/2.0 200 OK
Via: SIP/2.0/UDP scscf.homenet.com;branch=z9hG4bKhj5fy
```

```
Via: SIP/2.0/UDP pcscf.visitednet.com;branch=z9hG4bK7866
Via: SIP/2.0/UDP [::0000:1111:2222:3333]:2343;comp=sigcomp
 ;branch=z9hG4bK454245
To: <tel:+1-314-555-1234>;tag=g41334
From: <sip:mobile1@homenet.com>;tag=34199357834
P-Access-Network-Info: 3GPP-UTRAN-TDD
 ;utran-cell-id-3gpp=544542332
Call-ID: 7o5452gfvtyrfghdgrtt6900gh
CSeq: 16 PRACK
Content-Length: 0


SIP/2.0 200 OK
Via: SIP/2.0/UDP pcscf.visitednet.com;branch=z9hG4bK7866
Via: SIP/2.0/UDP [::0000:1111:2222:3333]:2343;comp=sigcomp
 ;branch=z9hG4bK454245
To: <tel:+1-314-555-1234>;tag=g41334
From: <sip:mobile1@homenet.com>;tag=34199357834
Call-ID: 7o5452gfvtyrfghdgrtt6900gh
CSeq: 16 PRACK
Content-Length: 0


SIP/2.0 200 OK
Via: SIP/2.0/UDP [::0000:1111:2222:3333]:2343;comp=sigcomp
 ;branch=z9hG4bK454245
To: <tel:+1-314-555-1234>;tag=g41334
From: <sip:mobile1@homenet.com>;tag=34199357834
Call-ID: 7o5452gfvtyrfghdgrtt6900gh
CSeq: 16 PRACK
Content-Length: 0


UPDATE sip:[::0000:aaaa:bbbb:cccc]:8767;comp=sigcomp SIP/2.0
Via: SIP/2.0/UDP [::0000:1111:2222:3333]:2343;comp=sigcomp
 ;branch=z9hG4bKkuyrt
Max-Forwards: 70
Route: <sip:pcscf.visitednet.com;lr>, <scscf.homenet.com;lr>,
 <sip:pscsf.homenet.com;lr>
To: <tel:+1-314-555-1234>;tag=g41334
From: <sip:mobile1@homenet.com>;tag=34199357834
P-Access-Network-Info: . . .
Call-ID: 7o5452gfvtyrfghdgrtt6900gh
CSeq: 17 UPDATE
Contact: <sip:[::0000:1111:2222:3333]:2343;comp=sigcomp>
Supported: 100rel
Require: precondition
Proxy-Require: sec-agree
Allow: INVITE, ACK, CANCEL, BYE, PRACK, MESSSAGE, REFER, UPDATE
Security-Verify: ipsec-3gpp; q=0.1; alg=hmac-sha-1-96; spi=34341
 ;port1=4323
Content-Type: application/sdp
Content-Length: . . .
```

```
v=0
o=- 2890844526 2890844527 IN IP6 [::0000:1111:2222:3333]
s=Phone Call
t=0 0
c=IN IP6 [::0000:1111:2222:3333]
m=audio 49170 RTP/AVP 97 96
b=AS:25.4
a=curr:qos local sendrecv
a=curr:qos remote none
a=des:qos mandatory local sendrecv
a=des:qos none remote sendrecv
a=rtpmap:97 AMR
a=fmtp:97 mode-set=0,2,5,7 ;maxframes=2
a=rtpmap:96 telephone-event
```

```
UPDATE sip:[::0000:aaaa:bbbb:cccc]:8767;comp=sigcomp SIP/2.0
Via: SIP/2.0/UDP pcscf.visitednet.com;branch=z9hG4bK56565
Via: SIP/2.0/UDP [::0000:1111:2222:3333]:2343;comp=sigcomp
 ;branch=z9hG4bKkuyrt
Max-Forwards: 69
Route: <scscf.homenet.com;lr>, <sip:pscsf.homenet.com;lr>
To: <tel:+1-314-555-1234>;tag=g41334
From: <sip:mobile1@homenet.com>;tag=34199357834
P-Access-Network-Info: . . .
P-Charging-Info: . . .
Call-ID: 7o5452gfvtyrfghdgrtt6900gh
CSeq: 17 UPDATE
Contact: <sip:[::0000:1111:2222:3333]:2343;comp=sigcomp>
Supported: 100rel
Require: precondition
Allow: INVITE, ACK, CANCEL, BYE, PRACK, MESSSAGE, REFER, UPDATE
Content-Type: application/sdp
Content-Length: ...
```

```
v=0
o=- 2890844526 2890844527 IN IP6 [::0000:1111:2222:3333]
s=Phone Call
t=0 0
c=IN IP6 [::0000:1111:2222:3333]
m=audio 49170 RTP/AVP 97 96
b=AS:25.4
a=curr:qos local sendrecv
a=curr:qos remote none
a=des:qos mandatory local sendrecv
a=des:qos none remote sendrecv
a=rtpmap:97 AMR
a=fmtp:97 mode-set=0,2,5,7 ;maxframes=2
a=rtpmap:96 telephone-event
```

```
UPDATE sip:[::0000:aaaa:bbbb:cccc]:8767;comp=sigcomp SIP/2.0
```

```
Via: SIP/2.0/UDP scscf.homenet.com;branch=z9hG4bKsa6wee
Via: SIP/2.0/UDP pcscf.visitednet.com;branch=z9hG4bK56565
Via: SIP/2.0/UDP [::0000:1111:2222:3333]:2343;comp=sigcomp
 ;branch=z9hG4bKkuyrt
Max-Forwards: 68
Route: <sip:pscsf.homenet.com;lr>
To: <tel:+1-314-555-1234>;tag=g41334
From: <sip:mobile1@homenet.com>;tag=34199357834
Call-ID: 7o5452gfvtyrfghdgrtt6900gh
CSeq: 17 UPDATE
Contact: <sip:[::0000:1111:2222:3333]:2343;comp=sigcomp>
Supported: 100rel
Allow: INVITE, ACK, CANCEL, BYE, PRACK, MESSSAGE, REFER, UPDATE
Content-Type: application/sdp
Content-Length: ...

v=0
o=- 2890844526 2890844527 IN IP6 [::0000:1111:2222:3333]
s=Phone Call
t=0 0
c=IN IP6 [::0000:1111:2222:3333]
m=audio 49170 RTP/AVP 97 96
b=AS:25.4
a=curr:qos local sendrecv
a=curr:qos remote none
a=des:qos mandatory local sendrecv
a=des:qos none remote sendrecv
a=rtpmap:97 AMR
a=fmtp:97 mode-set=0,2,5,7 ;maxframes=2
a=rtpmap:96 telephone-event


UPDATE sip:[::0000:aaaa:bbbb:cccc]:8767;comp=sigcomp SIP/2.0
Via: SIP/2.0/UDP pcscf.homenet.com;branch=z9hG4bK4gnm86
Via: SIP/2.0/UDP scscf.homenet.com;branch=z9hG4bKsa6wee
Via: SIP/2.0/UDP pcscf.visitednet.com;branch=z9hG4bK56565
Via: SIP/2.0/UDP [::0000:1111:2222:3333]:2343;comp=sigcomp
 ;branch=z9hG4bKkuyrt
Max-Forwards: 67
To: <tel:+1-314-555-1234>;tag=g41334
From: <sip:mobile1@homenet.com>;tag=34199357834
Call-ID: 7o5452gfvtyrfghdgrtt6900gh
CSeq: 17 UPDATE
Contact: <sip:[::0000:1111:2222:3333]:2343;comp=sigcomp>
Supported: 100rel
Require: precondition
Allow: INVITE, ACK, CANCEL, BYE, PRACK, MESSSAGE, REFER, UPDATE
Content-Type: application/sdp
Content-Length: ...

v=0
o=- 2890844526 2890844527 IN IP6 [::0000:1111:2222:3333]
s=Phone Call
t=0 0
```

```
c=IN IP6 [::0000:1111:2222:3333]
m=audio 49170 RTP/AVP 97 96
b=AS:25.4
a=curr:qos local sendrecv
a=curr:qos remote none
a=des:qos mandatory local sendrecv
a=des:qos none remote sendrecv
a=rtpmap:97 AMR
a=fmtp:97 mode-set=0,2,5,7 ;maxframes=2
a=rtpmap:96 telephone-event


SIP/2.0 200 OK
Via: SIP/2.0/UDP pcscf.homenet.com;branch=z9hG4bK4gnm86
Via: SIP/2.0/UDP scscf.homenet.com;branch=z9hG4bKsa6wee
Via: SIP/2.0/UDP pcscf.visitednet.com;branch=z9hG4bK56565
Via: SIP/2.0/UDP [::0000:1111:2222:3333]:2343;comp=sigcomp
 ;branch=z9hG4bKkuyrt
To: <tel:+1-314-555-1234>;tag=g41334
From: <sip:mobile1@homenet.com>;tag=34199357834
Call-ID: 7o5452gfvtyrfghdgrtt6900gh
CSeq: 17 UPDATE
Content-Type: application/sdp
Content-Length: ...

v=0
o=- 28908445761 28908445762 IN IP6 [::0000:aaaa:bbbb:cccc]
s=-
t=0 0
c=IN IP6 [::0000:aaaa:bbbb:cccc]
m=audio 5434 RTP/AVP 97 96
b=AS:25.4
a=curr:qos local sendrecv
a=curr:qos remote sendrecv
a=des:qos mandatory local sendrecv
a=des:qos none remote sendrecv
a=rtpmap:97 AMR
a=fmtp:97 mode-set=0,2,5,7 ;maxframes=2
a=rtpmap:96 telephone-event


SIP/2.0 200 OK
Via: SIP/2.0/UDP scscf.homenet.com;branch=z9hG4bKsa6wee
Via: SIP/2.0/UDP pcscf.visitednet.com;branch=z9hG4bK56565
Via: SIP/2.0/UDP [::0000:1111:2222:3333]:2343;comp=sigcomp
 ;branch=z9hG4bKkuyrt
To: <tel:+1-314-555-1234>;tag=g41334
From: <sip:mobile1@homenet.com>;tag=34199357834
Call-ID: 7o5452gfvtyrfghdgrtt6900gh
CSeq: 17 UPDATE
Content-Type: application/sdp
Content-Length: ...

v=0
```

```
o=- 28908445761 28908445762 IN IP6 [::0000:aaaa:bbbb:cccc]
s=-
t=0 0
c=IN IP6 [::0000:aaaa:bbbb:cccc]
m=audio 5434 RTP/AVP 97 96
b=AS:25.4
a=curr:qos local sendrecv
a=curr:qos remote sendrecv
a=des:qos mandatory local sendrecv
a=des:qos none remote sendrecv
a=rtpmap:97 AMR
a=fmtp:97 mode-set=0,2,5,7 ;maxframes=2
a=rtpmap:96 telephone-event


SIP/2.0 200 OK
Via: SIP/2.0/UDP pcscf.visitednet.com;branch=z9hG4bK56565
Via: SIP/2.0/UDP [::0000:1111:2222:3333]:2343;comp=sigcomp
 ;branch=z9hG4bKkuyrt
To: <tel:+1-314-555-1234>;tag=g41334
From: <sip:mobile1@homenet.com>;tag=34199357834
Call-ID: 7o5452gfvtyrfghdgrtt6900gh
CSeq: 17 UPDATE
Content-Type: application/sdp
Content-Length: ...

v=0
o=- 28908445761 28908445762 IN IP6 [::0000:aaaa:bbbb:cccc]
s=-
t=0 0
c=IN IP6 [::0000:aaaa:bbbb:cccc]
m=audio 5434 RTP/AVP 97 96
b=AS:25.4
a=curr:qos local sendrecv
a=curr:qos remote sendrecv
a=des:qos mandatory local sendrecv
a=des:qos none remote sendrecv
a=rtpmap:97 AMR
a=fmtp:97 mode-set=0,2,5,7 ;maxframes=2
a=rtpmap:96 telephone-event


SIP/2.0 200 OK
Via: SIP/2.0/UDP [::0000:1111:2222:3333]:2343;comp=sigcomp
 ;branch=z9hG4bKkuyrt
To: <tel:+1-314-555-1234>;tag=g41334
From: <sip:mobile1@homenet.com>;tag=34199357834
Call-ID: 7o5452gfvtyrfghdgrtt6900gh
CSeq: 17 UPDATE
Content-Type: application/sdp
Content-Length: ...

v=0
o=- 28908445761 28908445762 IN IP6 [::0000:aaaa:bbbb:cccc]
```

```
s=-
t=0 0
c=IN IP6 [::0000:aaaa:bbbb:cccc]
m=audio 5434 RTP/AVP 97 96
b=AS:25.4
a=curr:qos local sendrecv
a=curr:qos remote sendrecv
a=des:qos mandatory local sendrecv
a=des:qos none remote sendrecv
a=rtpmap:97 AMR
a=fmtp:97 mode-set=0,2,5,7 ;maxframes=2
a=rtpmap:96 telephone-event
```

```
SIP/2.0 180 Ringing
Via: SIP/2.0/UDP pcscf.homenet.com;branch=z9hG4bKa3423
Via: SIP/2.0/UDP scscf.homenet.com;branch=z9hG4bKkut
Via: SIP/2.0/UDP pcscf.visitednet.com;branch=z9hG4bKghf
Via: SIP/2.0/UDP [::0000:1111:2222:3333]:2343;comp=sigcomp
 ;branch=z9hG4bKdfewqh
Record-Route: <sip:pscsf.homenet.com;lr>, <scscf.homenet.com;lr>,
 <sip:pcscf.visitednet.com;lr>
To: <tel:+1-314-555-1234>;tag=g41334
From: <sip:mobile1@homenet.com>;tag=34199357834
Call-ID: 7o5452gfvtyrfghdgrtt6900gh
CSeq: 15 INVITE
Contact: <sip:[::0000:aaaa:bbbb:cccc]:8767;comp=sigcomp>
Require: 100rel
Allow: INVITE, ACK, CANCEL, BYE, PRACK, MESSSAGE, REFER, UPDATE
Content-Length: 0
```

```
SIP/2.0 180 Ringing
Via: SIP/2.0/UDP scscf.homenet.com;branch=z9hG4bKkut
Via: SIP/2.0/UDP pcscf.visitednet.com;branch=z9hG4bKghf
Via: SIP/2.0/UDP [::0000:1111:2222:3333]:2343;comp=sigcomp
 ;branch=z9hG4bKdfewqh
Record-Route: <sip:pscsf.homenet.com;lr>, <scscf.homenet.com;lr>,
 <sip:pcscf.visitednet.com;lr>
To: <tel:+1-314-555-1234>;tag=g41334
From: <sip:mobile1@homenet.com>;tag=34199357834
Call-ID: 7o5452gfvtyrfghdgrtt6900gh
CSeq: 15 INVITE
Contact: <sip:[::0000:aaaa:bbbb:cccc]:8767;comp=sigcomp>
Require: 100rel
Allow: INVITE, ACK, CANCEL, BYE, PRACK, MESSSAGE, REFER, UPDATE
Content-Length: 0
```

```
SIP/2.0 180 Ringing
Via: SIP/2.0/UDP pcscf.visitednet.com;branch=z9hG4bKghf
Via: SIP/2.0/UDP [::0000:1111:2222:3333]:2343;comp=sigcomp
 ;branch=z9hG4bKdfewqh
Record-Route: <sip:pscsf.homenet.com;lr>, <scscf.homenet.com;lr>,
```

```
  <sip:pcscf.visitednet.com;lr>
To: <tel:+1-314-555-1234>;tag=g41334
From: <sip:mobile1@homenet.com>;tag=34199357834
Call-ID: 7o5452gfvtyrfghdgrtt6900gh
CSeq: 15 INVITE
Contact: <sip:[::0000:aaaa:bbbb:cccc]:8767;comp=sigcomp>
Require: 100rel
Allow: INVITE, ACK, CANCEL, BYE, PRACK, MESSSAGE, REFER, UPDATE
Content-Length: 0


SIP/2.0 180 Ringing
Via: SIP/2.0/UDP [::0000:1111:2222:3333]:2343;comp=sigcomp
 ;branch=z9hG4bKdfewqh
Record-Route: <sip:pscsf.homenet.com;lr>, <scscf.homenet.com;lr>,
 <sip:pcscf.visitednet.com;lr>
To: <tel:+1-314-555-1234>;tag=g41334
From: <sip:mobile1@homenet.com>;tag=34199357834
Call-ID: 7o5452gfvtyrfghdgrtt6900gh
CSeq: 15 INVITE
Contact: <sip:[::0000:aaaa:bbbb:cccc]:8767;comp=sigcomp>
Require: 100rel
Allow: INVITE, ACK, CANCEL, BYE, PRACK, MESSSAGE, REFER, UPDATE
Content-Length: 0


SIP/2.0 200 OK
Via: SIP/2.0/UDP pcscf.homenet.com;branch=z9hG4bKa3423
Via: SIP/2.0/UDP scscf.homenet.com;branch=z9hG4bKkut
Via: SIP/2.0/UDP pcscf.visitednet.com;branch=z9hG4bKghf
Via: SIP/2.0/UDP [::0000:1111:2222:3333]:2343;comp=sigcomp
 ;branch=z9hG4bKdfewqh
Record-Route: <sip:pscsf.homenet.com;lr>, <scscf.homenet.com;lr>,
 <sip:pcscf.visitednet.com;lr>
To: <tel:+1-314-555-1234>;tag=g41334
From: <sip:mobile1@homenet.com>;tag=34199357834
Call-ID: 7o5452gfvtyrfghdgrtt6900gh
CSeq: 15 INVITE
Contact: <sip:[::0000:aaaa:bbbb:cccc]:8767;comp=sigcomp>
Require: 100rel
Allow: INVITE, ACK, CANCEL, BYE, PRACK, MESSSAGE, REFER, UPDATE
Content-Length: 0


SIP/2.0 200 OK
Via: SIP/2.0/UDP scscf.homenet.com;branch=z9hG4bKkut
Via: SIP/2.0/UDP pcscf.visitednet.com;branch=z9hG4bKghf
Via: SIP/2.0/UDP [::0000:1111:2222:3333]:2343;comp=sigcomp
 ;branch=z9hG4bKdfewqh
Record-Route: <sip:pscsf.homenet.com;lr>, <scscf.homenet.com;lr>,
 <sip:pcscf.visitednet.com;lr>
To: <tel:+1-314-555-1234>;tag=g41334
From: <sip:mobile1@homenet.com>;tag=34199357834
Call-ID: 7o5452gfvtyrfghdgrtt6900gh
```

```
CSeq: 15 INVITE
Contact: <sip:[::0000:aaaa:bbbb:cccc]:8767;comp=sigcomp>
Require: 100rel
Allow: INVITE, ACK, CANCEL, BYE, PRACK, MESSSAGE, REFER, UPDATE
Content-Length: 0


SIP/2.0 200 OK
Via: SIP/2.0/UDP pcscf.visitednet.com;branch=z9hG4bKghf
Via: SIP/2.0/UDP [::0000:1111:2222:3333]:2343;comp=sigcomp
 ;branch=z9hG4bKdfewqh
Record-Route: <sip:pscsf.homenet.com;lr>, <scscf.homenet.com;lr>,
 <sip:pcscf.visitednet.com;lr>
To: <tel:+1-314-555-1234>;tag=g41334
From: <sip:mobile1@homenet.com>;tag=34199357834
Call-ID: 7o5452gfvtyrfghdgrtt6900gh
CSeq: 15 INVITE
Contact: <sip:[::0000:aaaa:bbbb:cccc]:8767;comp=sigcomp>
Require: 100rel
Allow: INVITE, ACK, CANCEL, BYE, PRACK, MESSSAGE, REFER, UPDATE
Content-Length: 0


SIP/2.0 200 OK
Via: SIP/2.0/UDP [::0000:1111:2222:3333]:2343;comp=sigcomp
 ;branch=z9hG4bKdfewqh
Record-Route: <sip:pscsf.homenet.com;lr>, <scscf.homenet.com;lr>,
 <sip:pcscf.visitednet.com;lr>
To: <tel:+1-314-555-1234>;tag=g41334
From: <sip:mobile1@homenet.com>;tag=34199357834
Call-ID: 7o5452gfvtyrfghdgrtt6900gh
CSeq: 15 INVITE
Contact: <sip:[::0000:aaaa:bbbb:cccc]:8767;comp=sigcomp>
Require: 100rel
Allow: INVITE, ACK, CANCEL, BYE, PRACK, MESSSAGE, REFER, UPDATE
Content-Length: 0


ACK sip:[::0000:aaaa:bbbb:cccc]:8767;comp=sigcomp SIP/2.0
Via: SIP/2.0/UDP [::0000:1111:2222:3333]:2343;comp=sigcomp
 ;branch=z9hG4bK54343
Route: <sip:pcscf.visitednet.com;lr>, <scscf.homenet.com;lr>,
 <sip:pscsf.homenet.com;lr>
Max-Forwards: 70
To: <tel:+1-314-555-1234>;tag=g41334
From: <sip:mobile1@homenet.com>;tag=34199357834
P-Access-Network-Info: . . .
Call-ID: 7o5452gfvtyrfghdgrtt6900gh
CSeq: 15 ACK
Contact: <sip:[::0000:1111:2222:3333]:2343;comp=sigcomp>
Allow: INVITE, ACK, CANCEL, BYE, PRACK, MESSSAGE, REFER, UPDATE
Content-Length: 0
```

```
ACK sip:[::0000:aaaa:bbbb:cccc]:8767;comp=sigcomp SIP/2.0
Via: SIP/2.0/UDP pcscf.visitednet.com;branch=z9hG4bKnv3
Via: SIP/2.0/UDP [::0000:1111:2222:3333]:2343;comp=sigcomp
 ;branch=z9hG4bK54343
Route: <scscf.homenet.com;lr>,
 <sip:pscsf.homenet.com;lr>
Max-Forwards: 69
To: <tel:+1-314-555-1234>;tag=g41334
From: <sip:mobile1@homenet.com>;tag=34199357834
P-Access-Network-Info: 394893842917847375
Call-ID: 7o5452gfvtyrfghdgrtt6900gh
CSeq: 15 ACK
Contact: <sip:[::0000:1111:2222:3333]:2343;comp=sigcomp>
Allow: INVITE, ACK, CANCEL, BYE, PRACK, MESSSAGE, REFER, UPDATE
Content-Length: 0


ACK sip:[::0000:aaaa:bbbb:cccc]:8767;comp=sigcomp SIP/2.0
Via: SIP/2.0/UDP scscf.homenet.com;branch=z9hG4bK6fgn2
Via: SIP/2.0/UDP pcscf.visitednet.com;branch=z9hG4bKnv3
Via: SIP/2.0/UDP [::0000:1111:2222:3333]:2343;comp=sigcomp
 ;branch=z9hG4bK54343
Route: <sip:pscsf.homenet.com;lr>
Max-Forwards: 68
To: <tel:+1-314-555-1234>;tag=g41334
From: <sip:mobile1@homenet.com>;tag=34199357834
P-Access-Network-Info: . . .
Call-ID: 7o5452gfvtyrfghdgrtt6900gh
CSeq: 15 ACK
Contact: <sip:[::0000:1111:2222:3333]:2343;comp=sigcomp>
Allow: INVITE, ACK, CANCEL, BYE, PRACK, MESSSAGE, REFER, UPDATE
Content-Length: 0


ACK sip:[::0000:aaaa:bbbb:cccc]:8767;comp=sigcomp SIP/2.0
Via: SIP/2.0/UDP pcscf.homenet.com;branch=z9hG4bK45hmk
Via: SIP/2.0/UDP scscf.homenet.com;branch=z9hG4bK6fgn2
Via: SIP/2.0/UDP pcscf.visitednet.com;branch=z9hG4bKnv3
Via: SIP/2.0/UDP [::0000:1111:2222:3333]:2343;comp=sigcomp
 ;branch=z9hG4bK54343
Max-Forwards: 67
To: <tel:+1-314-555-1234>;tag=g41334
From: <sip:mobile1@homenet.com>;tag=34199357834
P-Access-Network-Info: . . .
Call-ID: 7o5452gfvtyrfghdgrtt6900gh
CSeq: 15 ACK
Contact: <sip:[::0000:1111:2222:3333]:2343;comp=sigcomp>
Allow: INVITE, ACK, CANCEL, BYE, PRACK, MESSSAGE, REFER, UPDATE
Content-Length: 0
```

10.8   Call Setup Example with Two Proxies

This section contains the complete message flow shown in Figure 2.2.

```
M1    INVITE sip:werner.heisenberg@munich.de SIP/2.0
      Via: SIP/2.0/UDP 100.101.102.103:5060;branch=z9hG4bKmp17a
      Max-Forwards: 70
      To: Heisenberg <sip:werner.heisenberg@munich.de>
      From: E. Schroedinger <sip:schroed5244@aol.com>;tag=42
      Call-ID: 10@100.101.102.103
      CSeq: 1 INVITE
      Subject: Where are you exactly?
      Contact: <sip:schroed5244@pc33.aol.com>
      Content-Type: application/sdp
      Content-Length: 159

      v=0
      o=schroed5244 2890844526 2890844526 IN IP4 100.101.102.103
      s=Phone Call
      t=0 0
      c=IN IP4 100.101.102.103
      m=audio 49170 RTP/AVP 0
      a=rtpmap:0 PCMU/8000


M2    INVITE sip:werner.heisenberg@200.201.202.203 SIP/2.0
      Via: SIP/2.0/UDP proxy.munich.de:5060;branch=z9hG4bK83842.1
      Via: SIP/2.0/UDP 100.101.102.103:5060;branch=z9hG4bKmp17a
      Max-Forwards: 69
      To: Heisenberg <sip:werner.heisenberg@munich.de>
      From: E. Schroedinger <sip:schroed5244@aol.com>;tag=42
      Call-ID: 10@100.101.102.103
      CSeq: 1 INVITE
      Contact: <sip:schroed5244@pc33.aol.com>
      Content-Type: application/sdp
      Content-Length: 159

      v=0
      o=schroed5244 2890844526 2890844526 IN IP4 100.101.102.103
      s=Phone Call
      c=IN IP4 100.101.102.103
      t=0 0
      m=audio 49172 RTP/AVP 0
      a=rtpmap:0 PCMU/8000


M3    SIP/2.0 180 Ringing
      Via: SIP/2.0/UDP proxy.munich.de:5060;branch=z9hG4bK83842.1
       ;received=100.101.102.105
      Via: SIP/2.0/UDP 100.101.102.103:5060;branch=z9hG4bKmp17a
      To: Heisenberg <sip:werner.heisenberg@munich.de>;tag=314159
      From: E. Schroedinger <sip:schroed5244@aol.com>;tag=42
      Call-ID: 10@100.101.102.103
      CSeq: 1 INVITE
```

```
     Contact: <sip:werner.heisenberg@200.201.202.203>
     Content-Length: 0


M4   SIP/2.0 180 Ringing
     Via: SIP/2.0/UDP 100.101.102.103:5060;branch=z9hG4bKmp17a
     To: Heisenberg <sip:werner.heisenberg@munich.de>;tag=314159
     From: E. Schroedinger <sip:schroed5244@aol.com>;tag=42
     Call-ID: 10@100.101.102.103
     CSeq: 1 INVITE
     Contact: <sip:werner.heisenberg@200.201.202.203>
     Content-Length: 0


M5   SIP/2.0 200 OK
     Via: SIP/2.0/UDP proxy.munich.de:5060;branch=z9hG4bK83842.1
       ;received=100.101.102.105
     Via: SIP/2.0/UDP 100.101.102.103:5060;branch=z9hG4bKmp17a
     To: Heisenberg7 <sip:werner.heisenberg@munich.de>;tag=314159
     From: E. Schroedinger <sip:schroed5244@aol.com>;tag=42
     Call-ID: 10@100.101.102.103
     CSeq: 1 INVITE
     Contact: <sip:werner.heisenberg@200.201.202.203>
     Content-Type: application/sdp
     Content-Length: 159

     v=0
     o=heisenberg 2890844526 2890844526 IN IP4 200.201.202.203
     s=Phone Call
     c=IN IP4 200.201.202.203
     t=0 0
     m=audio 49172 RTP/AVP 0
     a=rtpmap:0 PCMU/8000


M6   SIP/2.0 200 OK
     Via: SIP/2.0/UDP 100.101.102.103:5060;branch=z9hG4bKmp17a
     To: Heisenberg <sip:werner.heisenberg@munich.de>;tag=314159
     From: E. Schroedinger <sip:schroed5244@aol.com>;tag=42
     Call-ID: 10@100.101.102.103
     CSeq: 1 INVITE
     Contact: sip:werner.heisenberg@200.201.202.203
     Content-Type: application/sdp
     Content-Length: 159

     v=0
     o=heisenberg 2890844526 2890844526 IN IP4 200.201.202.203
     c=IN IP4 200.201.202.203
     t=0 0
     m=audio 49170 RTP/AVP 0
     a=rtpmap:0 PCMU/8000


M7   ACK sip:werner.heisenberg@200.201.202.203 SIP/2.0
```

```
        Via: SIP/2.0/UDP 100.101.102.103:5060;branch=z9hG4bKka42
        Max-Forwards: 70
        To: Heisenberg <sip:werner.heisenberg@munich.de>;tag=314159
        From: E. Schroedinger <sip:schroed5244@aol.com>;tag=42
        Call-ID: 10@100.101.102.103
        CSeq: 1 ACK
        Content-Length: 0


M8      BYE sip:schroed5244@pc33.aol.com SIP/2.0
        Via: SIP/2.0/UDP 200.201.202.203:5060;branch=z9hG4bK4332
        Max-Forwards: 70
        To: E. Schroedinger <sip:schroed5244@aol.com>;tag=42
        From: Heisenberg <sip:werner.heisenberg@munich.de>;tag=314159
        Call-ID: 10@100.101.102.103
        CSeq: 2000 BYE
        Content-Length: 0


M9      SIP/2.0 200 OK
        Via: SIP/2.0/UDP 200.201.202.203:5060;branch=z9hG4bK4332
        To: E. Schroedinger <sip:schroed5244@aol.com>;tag=42
        From: Heisenberg <sip:werner.heisenberg@munich.de>;tag=314159
        Call-ID: 10@100.101.102.103
        CSeq: 2000 BYE
        Content-Length: 0
```

10.9   SIP Presence and Instant Message Example

This section contains the call flow details of Figure 2.4.

```
M1      SUBSCRIBE sip:poisson@probability.org SIP/2.0
        Via SIP/2.0/TCP lecturehall21.academy.ru:5060
          ;branch=z9hG4bK348471123
        Max-Forwards: 70
        To: M. Poisson <sip:poisson@probability.org>
        From: P. L. Chebychev <sip:chebychev@academy.ru>;tag=21171
        Call-ID: 58dkfj349241k34452k592520
        CSeq: 3412 SUBSCRIBE
        Allow-Events: presence
        Allow: ACK, INVITE, CANCEL, BYE, NOTIFY, SUBSCRIBE, MESSAGE
        Contact: <sip:pafnuty@lecturehall21.academy.ru;transport=tcp>
        Event: presence
        Content-Length: 0


M2      SIP/2.0 202 Accepted
        Via SIP/2.0/TCP lecturehall21.academy.ru:5060
          ;branch=z9hG4bK348471123;received=19.34.3.1
        To: M. Poisson <sip:poisson@probability.org>;tag=25140
        From: P. L. Chebychev <sip:chebychev@academy.ru>;tag=21171
        Call-ID: 58dkfj349241k34452k592520
```

```
     CSeq: 3412 SUBSCRIBE
     Allow-Events: presence
     Allow: ACK, INVITE, CANCEL, BYE, NOTIFY, SUBSCRIBE, MESSAGE
     Contact: <sip:s.possion@dist.probability.org;transport=tcp>
     Event: presence
     Expires: 3600
     Content-Length: 0


M3   NOTIFY sip:pafnuty@lecturehall21.academy.ru SIP/2.0
     Via SIP/2.0/TCP dist.probablilty.org:5060
      ;branch=z9hG4bK4321
     Max-Forwards: 70
     To: P. L. Chebychev <sip:chebychev@academy.ru>;tag=21171
     From: M. Poisson <sip:poisson@probability.org>;tag=25140
     Call-ID: 58dkfj349241k34452k592520
     CSeq: 1026 NOTIFY
     Allow: ACK, INVITE, CANCEL, BYE, NOTIFY, SUBSCRIBE, MESSAGE
     Allow-Events: dialog
     Contact: <sip:s.possion@dist.probability.org;transport=tcp>
     Subscription-State: active;expires=3600
     Event: presence
     Content-Type: application/cpim-pidf+xml
     Content-Length: 244

     <?xml version="1.0" encoding="UTF-8"?>
     <presence xmlns="urn:ietf:params:xml:ns:cpim-pidf"
       entity="sip:poisson@probability.org">
      <tuple id="452426775">
       <status>
        <basic>closed</basic>
       </status>
      </tuple>
     </presence>


M4   SIP/2.0 200 OK
     Via SIP/2.0/TCP dist.probablilty.org:5060
      ;branch=z9hG4bK4321;received=24.32.1.3
     To: P. L. Chebychev <sip:chebychev@academy.ru>;tag=21171
     From: M. Poisson <sip:poisson@probability.org>;tag=25140
     Call-ID: 58dkfj349241k34452k592520
     CSeq: 1026 NOTIFY
     Content-Length: 0


M5   NOTIFY sip:pafnuty@lecturehall21.academy.ru SIP/2.0
     Via SIP/2.0/TCP dist.probablilty.org:5060
      ;branch=z9hG4bK334241
     Max-Forwards: 70
     To: P. L. Chebychev <sip:chebychev@academy.ru>;tag=21171
     From: M. Poisson <sip:poisson@probability.org>;tag=25140
     Call-ID: 58dkfj349241k34452k592520
     CSeq: 1027 NOTIFY
```

```
      Allow: ACK, INVITE, CANCEL, BYE, NOTIFY, SUBSCRIBE, MESSAGE
      Allow-Events: presence
      Contact: <sip:s.possion@dist.probability.org;transport=tcp>
      Subscription-State: active;expires=1800
      Event: presence
      Content-Type: application/cpim-pidf+xml
      Content-Length: 325

      <?xml version="1.0" encoding="UTF-8"?>
      <presence xmlns="urn:ietf:params:xml:ns:cpim-pidf"
        entity="sip:poisson@probability.org">
       <tuple id="452426775">
        <status>
          <basic>open</basic>
        </status>
          <contact>sip:s.possion@dist.probability.org;transport=tcp
          </contact>
       </tuple>
      </presence>


M6    SIP/2.0 200 OK
      Via SIP/2.0/TCP dist.probablilty.org:5060
        ;branch=z9hG4bK334241;received=24.32.1.3
      To: P. L. Chebychev <sip:chebychev@academy.ru>;tag=21171
      From: M. Poisson <sip:poisson@probability.org>;tag=25140
      Call-ID: 58dkfj349241k34452k592520
      CSeq: 1027 NOTIFY
      Content-Length: 0


M7    MESSAGE sip:s.possion@dist.probability.org SIP/2.0
      Via SIP/2.0/TCP lecturehall21.academy.ru:5060
        ;branch=z9hG4bK3gtr2
      Max-Forwards: 70
      To: M. Poisson <sip:s.possion@dist.probability.org>
      From: P. L. Chebychev <sip:chebychev@academy.ru>;tag=4542
      Call-ID: 9dkei93vjq1ei3
      CSeq: 15 MESSAGE
      Allow: ACK, INVITE, CANCEL, BYE, NOTIFY, SUBSCRIBE, MESSAGE
      Content-Type: text/plain
      Content-Length: 9

      Hi There!


M8    SIP/2.0 200 OK
      Via SIP/2.0/TCP lecturehall21.academy.ru:5060
        ;branch=z9hG4bK3gtr2;received=19.34.3.1
      To: M. Poisson <sip:s.possion@dist.probability.org>;tag=2321
      From: P. L. Chebychev <sip:chebychev@academy.ru>;tag=4542
      Call-ID: 9dkei93vjq1ei3
      CSeq: 15 MESSAGE
      Content-Length: 0
```

```
M9   MESSAGE sip:chebychev@academy.ru SIP/2.0
     Via SIP/2.0/TCP dist.probablilty.org:5060
      ;branch=z9hG4bK4526245
     Max-Forwards: 70
     To: P. L. Chebychev <sip:chebychev@academy.ru>
     From: M. Poisson <sip:s.possion@dist.probability.org>;tag=14083
     Call-ID: 1k34452k592520
     CSeq: 2321 MESSAGE
     Allow: ACK, INVITE, CANCEL, BYE, NOTIFY, SUBSCRIBE, MESSAGE
     Content-Type: text/plain
     Content-Length: 30

     Well, hello there to you, too!


M10  SIP/2.0 200 OK
     Via SIP/2.0/TCP dist.probablilty.org:5060
      ;branch=z9hG4bK4526245;received=24.32.1.3
     To: P. L. Chebychev <sip:chebychev@academy.ru>;tag=mc3bg5q77wms
     From: M. Poisson <sip:s.possion@dist.probability.org>;tag=14083
     Call-ID: 1k34452k592520
     CSeq: 2321 MESSAGE
     Content-Length: 0
```

References

[1]   Rosenberg, J., et al., "SIP: Session Initiation Protocol," RFC 3261, 2002.

[2]   Johnston, A., et al., "Basic SIP Call Flow Examples," IETF Internet-Draft, Work in Progress, April 2003.

[3]   Johnston, A., et al., "SIP/PSTN Call Flow Examples," IETF Internet-Draft, Work in Progress, April 2003.

[4]   Schulzrinne, H., and H. Agrawl, "Session Initiation Protocol (SIP)-H.323 Interworking Requirements," IETF Internet-Draft, Work in Progress, February 2003.

11

Future Directions

This chapter will discuss some future areas of work in SIP-related working groups in the IETF. Instead of attempting to list and discuss a snapshot of current activity in the IETF, the reader should gather the information directly from the IETF itself. Table 11.1 lists the charter Web pages for the three most important SIP working groups: SIP, SIPPING, and SIMPLE. The charter page for each working group lists the deliverables of the group along with RFCs (finished documents) and Internet-Drafts (works in progress). Only Internet-Drafts that have been adopted as official work group items are listed on these Web pages—these are the documents most likely to become RFCs in the near future. The Web page also contains information about joining the working group e-mail list, which discusses the listed set of Internet-Drafts.

Finally, one can search the IETF Internet-Draft archives for documents relating to SIP at http://search.ietf.org. However, be warned: There are many, many documents and most will likely never be published as an RFC—always consult someone familiar with the working group activity before assuming that an Internet-Draft not listed on a working group charter page is likely to ever become an Internet Standard.

11.1 SIP, SIPPING, and SIMPLE Working Group Design Teams

From time to time working groups use a design team to work on a particular topic. The teams report regularly to the working group and have a working group chair who provides guidance. The approach has been successful in the past and will likely result in new RFCs being produced by the design teams described in the next sections.

Table 11.1

Latest Status of SIP-Related IETF Work

Working Group	Web Page URL	Area
SIP	http://www.ietf.org/html.charters/sip-charter.html	SIP protocol extensions
SIPPING	http://www.ietf.org/html.charters/sipping-charter.html	SIP requirements, best current practice documents, and usage
SIMPLE	http://www.ietf.org/html.charters/simple-charter.html	SIP for IM and presence
XCON	http://www.ietf.org/html.charters/xcon-charter.html	Centralized conferencing

Some of the referenced Internet-Drafts in this chapter may have become RFCs or have changed. Useful archives of expired Internet-Drafts are available at http://www.softarmor.com/sipwg, http://www.softarmor.com/sipping, http://www.softarmor.com/simple, http://www.softarmor.com/xcon, and http://www.iptel.org/info/players/ietf.

11.1.1 SIP and Hearing Impairment Design Team

This design team is developing requirements, guidelines, and directions for SIP to support the needs of hearing impaired users. The design team has produced a requirements document [1] that includes scenarios and requirements. The work has also produced general requirements on transcoding, which have wider scope than the hearing impaired.

11.1.2 Conferencing Design Team

This design team is working on issues relating to tightly coupled multiparty conferencing. The work so far has produced a requirements [2], framework [3], and call control [4] documents that are SIPPING working group items. The framework includes voice, video, IM, and collaboration. Other items include the definition of the isfocus feature tag [5] and the conference event package [6].

The design team is also working on areas relating to standardizing XML messages for conference and media mixing policy. This will allow an intelligent end point or automata to set up and configure a multimedia conferencing session. The media policy manipulation will allow an end point to control their view and mixing of the overall conference. This work will be standardized in the new XCON (centralized conferencing) IETF working group. However, intelligent SIP end points and clients will likely support these extensions and protocols.

11.1.3 Application Interaction Design Team

This design team is developing a framework for application interaction using SIP. It will cover areas such as dual-tone multifrequency (DTMF) and stimulus (button pressing) transport and integration with SIP signaling.

11.1.4 Emergency Calling Design Team

This design team is developing requirements for SIP support of emergency calling (911 in the United States and 112 in other parts of the world).

11.2 Other SIP Work Areas

Some other important work areas relating to SIP are discussed in the following sections.

11.2.1 Emergency Preparedness

The IETF Internet Emergency Preparedness (IEPREP) working group has been working on requirements on how to make sure that the Internet functions properly and can provide reliable communications for emergency personnel in the event of an emergency or disaster. There is a document of requirements for SIP [7] that may produce some SIP extensions and conventions.

11.2.2 Globally Routable Contact URIs

Work is currently underway to try to provide a mechanism to guarantee that a `Contact` URI is globally routable. While in general, a `Contact` URI is globally routable, in the presence of NATs, firewalls, and screening proxies, this may not be so. Contributing to the problem is that a UA may not know if its `Contact` URI is globally routable or not. Current work is discussing the requirements, which, when agreed upon, can be translated into a SIP extension or convention to solve.

11.2.3 Service Examples

This service examples document [8] shows how intelligent end points can implement a set of service commonly found in PBX switches and in service provider Centrex offerings. The set of features includes call hold, call park, call pickup, auto recall, music on hold, attended and unattended transfer, and others. All of the flows use standard SIP operations and call control extensions.

11.3   SIP Instant Message and Presence Work

In the SIMPLE working group, there has been much work in the area of instant message sessions, similar to that shown in Figure 4.10 in which an `INVITE` sets up an IM session between the end points. The current approach being developed uses a SIP-like protocol called Message Session Relay Protocol (MSRP) [9], which allows the IM session to be established either directly end-to-end or through a relay necessary for logging or NAT/firewall traversal.

The SIMPLE working group is also working on documents to enable all aspects of a standard presence and IM client to be implemented using standard IETF protocols. While the basic ability is provided through the defined `MESSAGE`, `SUBSCRIBE`, and `NOTIFY` methods, tools to upload or publish presence information, and protocols for "buddy list" creation and manipulation are still needed.

Other interesting work continues in the area of "rich" presence information delivery using SIP [10]. The current presence XML formats only list contacts and describe them as "open" or "closed," essentially on-line or off-line. This work extends so that more presence information can be automatically updated based on calendars and other applications from the network.

References

[1]   Charlton, N., et al., "User Requirements for the Session Initiation Protocol (SIP) in Support of Deaf, Hard of Hearing and Speech-Impaired Individuals," RFC 3351, 2002.

[2]   Levin, O., and R. Evan, "High Level Requirements for Tightly Coupled SIP Conferencing," IETF Internet-Draft, Work in Progress, April 2003.

[3]   Rosenberg, J., "A Framework for Conferencing with the Session Initiation Protocol," IETF Internet-Draft, Work in Progress, May 2003.

[4]   Johnston, A., and O. Levin, "SIP Call Control – Conferencing for User Agents," IETF Internet-Draft, Work in Progress, April 2003.

[5]   Rosenberg, J., H. Schulzrinne, and P. Kyzivat, "Caller Preferences and Callee Capabilities for the Session Initiation Protocol (SIP)," IETF Internet-Draft, Work in Progress, March 2003.

[6]   Rosenberg, J., and H. Schulzrinne, "A Session Initiation Protocol (SIP) Event Package for Conference State," IETF Internet-Draft, Work in Progress, June 2002.

[7]   Peterson, J., "Considerations on the IEPREP Requirements for SIP," IETF Internet-Draft, Work in Progress, April 2003.

[8] Johnston, A., et al., "Session Initiation Protocol Service Examples," IETF Internet-Draft, Work in Progress, March 2003.

[9] Campbell, B., et al., "Instant Message Sessions in SIMPLE," IETF Internet-Draft, Work in Progress, April 2003.

[10] Schulzrinne, H., et al., "RPIDS—Rich Presence Information Data Format for Presence Based on the Session Initiation Protocol (SIP)," IETF Internet-Draft, Work in Progress.

Appendix A: Changes in the SIP Specification from RFC 2543 to RFC 3261

In late 2001 and early 2002, the base SIP specification RFC 2543 was rewritten and published as RFC 3261. The new RFC was backwards compatible with the old RFC except in areas in which the old RFC was deemed to be broken or buggy. Besides these fixes, the document was also rewritten from the ground up with a new structure and new coauthors to increase the clarity of the specification. The new structure included a redefinition of the SIP finite state machine (FSM) into three layers: transport, transaction, and transaction user (TU). Additional timers and time-outs were also defined. The core specification also had related topics such as SDP handling and DNS Procedures removed and published as separate specifications RFC 3264 and RFC 3263.

This appendix describes some of the key differences between the two RFCs. Many of these differences have been mentioned throughout the book—others have not been mentioned.

Some parts of RFC 2543 were removed or deprecated from RFC 3261. In the area of security, HTTP Basic authentication was removed, as it allowed the transport of passwords in the clear (HTTP Digest does not do this and is still supported in RFC 3261). Also deprecated was Pretty Good Privacy (PGP) encryption in favor of S/MIME encryption. PGP had not been implemented and the S/MIME approach is the one currently favored by the IETF to secure bodies end-to-end of the Internet. The Authentication header field (which was used for PGP encryption) was deprecated. The Response-Key and Hide header fields were also deprecated due to security problems associated with their use.

The new RFC also deprecates the use of Via header field loop detection in favor of a required Max-Forwards header field in every request. This is because of the difficulty and processing burden on proxies to parse long sets of Via headers.

Perhaps the most major change outside of security relates to the introduction of loose routing concepts in place of the strict routing used with Route headers in RFC 2543. In the old RFC, for example, a request containing three Route headers could only route to those three proxies then to the end point—no deviations were allowed. Also, a Route header could only be built from a received Record-Route header field. In RFC 3261, a request containing three loose Route headers must route through those three proxies but may also visit other proxies as needed. Also, a preloaded Route header may be constructed at the start of a dialog without first receiving a Record-Route header in dialog setup. This permits much more flexible routing for services and mobile applications. They can also be used to provide midcall services and home proxy services as discussed in Chapter 9. In addition, the mechanism used is much cleaner and understandable than the old "strict" routing, and it is also backwards compatible, even in a mixed network of strict and loose routing proxies.

A major improvement in RFC 3261 comes in the security area. Besides the deprecation of HTTP Basic authentication and PGP and the addition of S/MIME for bodies, the specification also introduces the secure SIP (sips) URI scheme, which requires the use of end-to-end TLS transport (with the exception of the last hop, which may use some other encryption mechanism). This allows a UA to know that the SIP message path does not traverse any unencrypted links end-to-end through the presence of intermediary proxies.

Other changes include a new requirement that Contact URIs be globally routable (which is useful in certain call control scenarios). SIP now also includes full support for IPv6 addresses. The use of hostnames instead of literal addresses is now recommended in Via and Contact URIs. The new specification renames the "call leg" as a dialog. The definition of a dialog was also changed so that it includes the To and From tags but not URIs. This will allow future versions of SIP to permit To and From URIs to be changed within a dialog for clarity. (For example, if A calls B but the call is forwarded to C, when C answers, the 200 OK could contain the URI of C instead of B in the To header field, alerting A that C not B has picked up the call.)

The new specification also formally defines an "early" dialog, which is created when a response to an INVITE (or SUSCRIBE) is received which contains a tag. In the case of forking, multiple independent early dialogs may be created for a single INVITE request. As a result, Contact and Record-Route header fields must be returned in responses that create early dialogs.

RFC 3261 also included new header fields such as `Alert-Info`, `Call-Info`, `Reply-To`, `In-Reply-To`, and `Error-Info`, which add new and useful functionality to SIP.

About the Author

Alan B. Johnston is a distinguished technical member with MCI in their engineering department, and is also an adjunct assistant professor of electrical and systems engineering at Washington University in St. Louis, Missouri. He is currently working with the SIP protocol in their MCI Advantage product. Prior to MCI, he worked at a competitive local exchange carrier (CLEC), Brooks Fiber Properties, Telcordia (formerly Bellcore), and SBC Technology Resources. Dr. Johnston is a coauthor of the new SIP specification RFC 3261 and the cochair of the new XCON Centralized Conferencing IETF Working Group. He has a Ph.D. from Lehigh University, Bethlehem, Pennsylvania, in electrical engineering and a B.S. in engineering (first class honors) from the University of Melbourne, Melbourne, Australia, in electrical engineering. Born in Melbourne, Australia, Dr. Johnston currently resides in a national historic district in St. Louis, Missouri, with his wife Lisa, son Aidan, and daughter Nora.

Index

Recent Titles in the Artech House Telecommunications Library

Vinton G. Cerf, Senior Series Editor

Telemetry Systems Engineering, Frank Carden, Russell Jedlicka, and Robert Henry

Telephone Switching Systems, Richard A. Thompson

Understanding Modern Telecommunications and the Information Superhighway, John G. Nellist and Elliott M. Gilbert

Understanding Networking Technology: Concepts, Terms, and Trends, Second Edition, Mark Norris

Videoconferencing and Videotelephony: Technology and Standards, Second Edition, Richard Schaphorst

Visual Telephony, Edward A. Daly and Kathleen J. Hansell

Wide-Area Data Network Performance Engineering, Robert G. Cole and Ravi Ramaswamy

Winning Telco Customers Using Marketing Databases, Rob Mattison

WLANs and WPANs towards 4G Wireless, Ramjee Prasad and Luis Muñoz

World-Class Telecommunications Service Development, Ellen P. Ward

For further information on these and other Artech House titles,
including previously considered out-of-print books now available through our
In-Print-Forever® (IPF®) program, contact:

Artech House
685 Canton Street
Norwood, MA 02062
Phone: 781-769-9750
Fax: 781-769-6334
e-mail: artech@artechhouse.com

Artech House
46 Gillingham Street
London SW1V 1AH UK
Phone: +44 (0)20 7596-8750
Fax: +44 (0)20 7630-0166
e-mail: artech-uk@artechhouse.com

Find us on the World Wide Web at:
www.artechhouse.com